No. 3097
$23.95

Wall Framing

Elizabeth and Robert Williams

TAB BOOKS Inc.
Blue Ridge Summit, PA

FIRST EDITION
FIRST PRINTING

Copyright © 1989 by TAB BOOKS Inc.
Printed in the United States of America

Library of Congress Cataloging in Publication Data

Williams, Elizabeth, 1942-
Wall framing / by Elizabeth and Robert Williams.
    p.   cm.
Includes index.
ISBN 0-8306-9097-2   ISBN 0-8306-9397-1 (pbk.)
1. Walls. 2. Framing (Building) I. Williams, Robert Leonard,
1932-   . II. Title.
TH2251.W55   1989
694'.2—dc19   88-32319
                                                  CIP

TAB BOOKS Inc. offers software for sale. For information and a catalog, please contact
TAB Software Department
Blue Ridge Summit, PA 17294-0850

Questions regarding the content of this book should be addressed to:

Reader Inquiry Branch
TAB BOOKS Inc.
Blue Ridge Summit, PA 17294-0214

Edited by Merry Stinson

Contents

Introduction

Although it is impossible to isolate a building's single most important structured element, the wall is certainly one of its most significant components. A good foundation is vital, as is a good roof, but whereas these elements each serve one basic purpose, the wall performs many roles.

Some walls support the ceiling and roof. These *load-bearing*, or *bearing* walls, must be durable, strong, and evenly constructed. Exterior bearing walls must support various sidings of aluminum, vinyl, wood, or masonry.

Other walls are *non-load-bearing*, or *nonbearing*: their major function is to partition. However, these walls, too, have additional purposes. Nonbearing walls must accommodate doorways, bookcases, closets, alcoves, fireplaces, wiring, and plumbing. These walls must also support paneling, sheathing, gypsum board or wallboard, wallpaper, fixtures, cabinets, and a wide variety of decorative and utilitarian appointments.

Both interior and exterior walls should be constructed so that they provide beauty, variety, insulation, and sound-proofing. They should act as a barrier to moisture, wind, extreme temperatures, and invasion by insects and other pests.

Properly constructed, walls offer a strong and durable surface on which the tenant might express himself creatively. Attractively decorated walls, enhanced by well-planned furnishings, create the unique character of a home.

Improperly constructed, walls cause frustration, concern, expense, and trouble. If walls are not squarely built, for instance, it is difficult to install paneling, wallpaper, or other wall coverings because the covering will not fit into corners. If studs are installed with improper spacing, there will be inadequate nailing surface for plasterboard.

Door framings incorrectly aligned cause doors to stick, sag, rattle, and refuse to close properly or stay closed. Doors hung on incorrect frames will gap badly at the top or bottom. Often they will scrape the floor and damage rugs and carpets, or leave unsightly scratches on hardwood floors. Exterior doors that are hung improperly permit insects, mice, moisture, cold air, and unnecessary noises to enter the house.

Misaligned windows might stick or rattle, leak costly heat or air conditioning, and cause difficulty in installing properly fitting storm windows. Poorly fitted windows allow rain to seep in around the frame and facing. This dampness can saturate insulation, moisten electrical wiring, and cause wood decay. This moisture also can cause severe warping of studding and clapboards, which results in serious damage to the house.

Moist walls become a haven for roaches, termites, and other insects. Ill-fitting windows also invite wasps, hornets, and other stinging creatures to nest inside the structure's dark recesses.

Uneven walls hinder the job of finishing ceilings and floors. If corners are not square and walls are not true, you will need to trim floor or ceiling tiles or shape them to fit the wall lines. If wall bracing is not installed properly, walls can bow, sag, and lose strength.

The only acceptable wall is one that is constructed properly. Anything less results in false economy and worry.

The purpose of this book is to assist the handyman, the serious hobbyist, the beginner carpenter, and the total novice in his efforts to construct wall framing that will endure, serve its purposes, and reward the efforts of the builder.

We wrote this book in layman's language when possible. We define technical terms in simple, easy-to-understand language, and give instructions in the simplest fashion.

We present information to enable one person, working without assistance, to handle any facet of wall framing. It is much easier if at least one other person helps hold heavy lengths of lumber while the carpenter nails them in place. It is not always feasible for a helper to be present, however, so this book includes numerous practical suggestions for solving problems unassisted.

We supply a list of necessary, helpful, and desirable tools. We address buying or renting some equipment, and offer several shortcuts that save time and effort without sacrificing quality.

We provide information on building permits, standard versus substandard building practices and materials, city or county inspections, building codes, and special licenses. We also explain installation of insulation, wiring, and plumbing. We tell how to straighten crooked studs and how to build crucial corner posts.

We do not provide step-by-step instructions on wiring and plumbing problems; however, we do offer suggestions that can simplify these tasks. We also address deviations from the usual wall structure such as closets and partial partitions. We provide helpful information on baseboards and molding.

We take the builder, whether a beginner or an experienced handyman, from the first steps in framing a wall to the completed frame that is ready for covering and decoration. We offer suggestions for dealing with problems that might arise.

This book is intended to be used, both before starting to work, and as an on-site guide. Therefore, as much information as possible is located with the pertinent operation.

In this book we provide several names for a product, operation, or structural part if the alternate labels clarify a step in the framing process. We define an archaic term at its first appearance in the text. We avoid using brand names for products.

You can save a great deal of money by framing your own walls because labor expenses represent about half the cost of a building project. You also have the satisfaction of knowing that the spare room or other project is basically your own work.

If you are patient, willing to stop to correct mistakes, and able to spare the time needed for careful work, you should have no real difficulty in framing a wall. Although wall framing is not easy, it can be a pleasant and worthwhile adventure if a few common techniques and basic guidelines are followed. The inexperienced carpenter can construct totally satisfactory walls by following the guidelines offered in this book.

1
Introduction
to Wall Framing

THE WALL FRAMES IN ANY HOUSE OR STRUCTURE ARE SIMILAR TO THE skeletal structure of the body. The caps, soles, and studs are the "bones" of the building and serve to keep it sturdy and strong. If the wall frames are perfect, then the walls themselves also can be durable and strong. If frames are poorly constructed, the entire house can become a series of problems.

If you are building a house, adding a room to your house, or partitioning an existing room, you will need to construct wall frames. Some of this framing will become part of a *bearing wall*. A bearing wall is one that supports, in part, the weight of all of the structure that is above it: ceiling, rafters, braces, trusses, roof sheathing, and roofing. In some instances, the bearing wall also must support part of the weight of rooms above it. Therefore, a bearing wall should be constructed as flawlessly as possible using the best materials and following basic building techniques.

A *nonbearing* wall supports only its own studding, braces, and windows and doors, if any. It does not carry any of the weight of the rest of the structure. The nonbearing wall should also be built carefully, however, because if it is built using inferior materials or workmanship, the wall can warp, sag, decay, or suffer damage from moisture or insects.

A professional maintenance man in North Carolina decided to put vinyl siding over his exterior walls, which, in addition to containing green lumber, were already slightly curved and moisture-damaged. Within a few weeks, the walls had warped and buckled so badly that the siding pulled loose and fell off. The entire job had to be redone, including correcting the defects in the existing walls.

Shortcuts in building practices, unless they are intelligent ones, all too frequently result in similar experiences. The proper building techniques might take a little longer, but they are much shorter, easier, and cheaper in the long run.

Erecting a wall frame that has square corners, perfectly spaced studs, sound soleplates and top plates, good corner posts, sound headers, and good cap plates is not a job that must be done only by experts. You can do it, too.

You can do it if you can use a rule or tape measure, drive a nail, saw a straight line, use a plumb bob, and perform a few basic carpentry functions. Chapter 2 discusses in detail the skills that are needed. First, however, you need to become familiar with the components of a good wall frame.

IMPORTANCE OF BEARING AND NONBEARING WALLS

The bearing wall rests on a foundation wall, which is supported by footings. The first step in house construction consists of digging trenches that outline the exterior of the entire building to a depth below the frost line. Several inches of coarse gravel are then laid over the soil, and concrete is poured over the gravel.

After the footings are allowed to set and piers of pillars are constructed, the foundation walls are added. A foundation wall is perhaps the most crucial of all the components of a house because it supports the entire house.

The *foundation sill*, a heavy beam, is placed on top of the foundation wall. (See Fig. 1-1.) The sill is the first part of the actual house framing to be set in place. Sills must be of the soundest lumber and must be protected against termites, water, and decay.

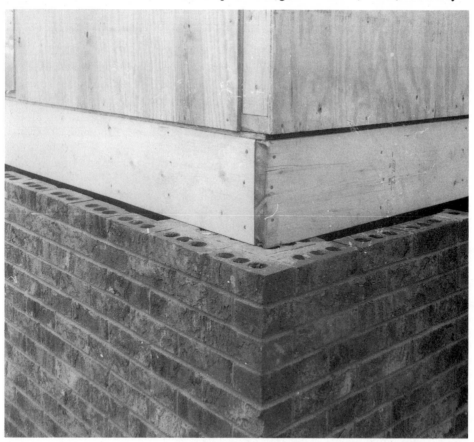

Fig. 1-1. If you plan to add a brick facing to the wall, leave enough space between sills and edge of foundation wall for sheathing.

Sometimes a *box sill structure* is used. This is simply a square or rectangle that follows the outline of the exterior of the house and rests upon the heavy lumber, called the *sill plate*. Sometimes it rests upon the foundation and is bolted in place.

Floor joists are then nailed in place. These are timbers that are usually 2 × 8 or 2 × 10 in dimension. They rest upon piers if their length is greater than 15 feet. Subflooring is nailed over the joists and box sill, and the flooring and wall framing are installed. (See Fig. 1-2.)

Fig. 1-2. For subflooring you can use 4-×-8 panels of plywood or similar material, or you can use tongue-and-groove or square-edged boards.

If the wall framing is defective and buckles, warps, or sags, cracks might form in the exterior siding or clapboard, admitting rain, wasps, hornets, termites, and other damaging insects and pests. In many houses, the window framing fits so poorly that rain can seep behind the frame, drip inside the window, run into the walls, soak the insulation, and eventually make its way to the bottom of the wall. If moisture causes the wood to rot and warp badly enough, bows and cracks large enough to admit mice may appear, and these rodents can gnaw enough wood to do irreparable damage to the structure.

Inside the wall, the water creates a perfect breeding place for termites, carpenter ants, and similar wood-devouring pests. Fungi appear and wood decay begins almost immediately. The subflooring, joists, box sill, and sill plates can decay in a very short time, and only a major repair job can salvage that part of the house.

Termites can tunnel their way completely through the studding and door frames so that only a very porous, paperlike shell is left of what had once been strong studding. Termites even reach the upstairs levels of houses, if the walls admit enough moisture for the pests' needs. When studs are attacked by fungus, termites, or decay, the entire wall sags first, then collapses.

As you can see, the bearing wall is more than just a support for the upper part of the house. It is the key to the endurance and life of the entire house.

The nonbearing wall does not serve the structural purpose that the bearing wall serves, but it is important for its decorative and partitioning qualities. Installed properly, it can add greatly to the beauty and service of a house. Installed in a less adequate fashion, it might be out of square or studding might be improperly spaced so that putting up gypsum board or other drywall materials, wallpaper, or paneling is subverted from a very simple and pleasant experience to an unforgettable and at times costly endeavor.

Thus, framing the wall properly from the beginning is essential. The way to get off to a good start is to learn the components of wall framing.

TERMS

Wall framing is made up of regular studs (sometimes called common studs), diagonal bracing, cripples (or cripple studs), stud blocks, trimmers, headers, soles or soleplates, corner posts, T-posts, top plates, double top plates, header joists, cut-in and let-in bracing, and window and door rough framing.

Common studs are usually 2-×-4 lengths of lumber that run vertically between the soleplate and top plate of a wall. Such studs, or *studding* as it is often called, are usually placed 16 inches apart on center, but in special conditions they can be set 12 or 24 inches apart. When the distance on center is greater than 16 inches, however, lumber larger than 2 × 4s is generally used. Studding provides excellent wall support.

Diagonal bracing is made up of 1-×-4 boards, or sometimes 1-×-6 boards, nailed to the soleplate about 4 feet from the corner, and to corner posts at a point about 4 feet high. Diagonal bracing strengthens wall framing by stiffening the wall and helping it to resist high winds, twisting, warping, and straining. Diagonal bracing also helps to keep corners square and walls perfectly vertical. If corners and walls are allowed to sag or buckle, plaster might crack and wallboard or paneling might warp and bow. Doors could stick, fit poorly, or sag so badly that they would not close well, and windows might either be too loose or so tight that they cannot be raised and lowered with ease.

Cripple studs, or *cripples*, are short studs that are nailed under and above windows and above door frames.

Stud blocks, or *blocking*, consist of short lengths of 2 × 4s nailed horizontally between studs to strengthen them and to keep them from curving or bowing. This blocking is usually nailed halfway up the studs and is staggered to permit easier nailing. The first blocking can be nailed 4 feet from the soleplate, and the second one can be placed so that the top is even with the bottom of the first one, or 2 inches lower. The third one then can be nailed in line with the first, and the fourth can be in line with the second. This pattern continues across the entire wall.

A *trimmer* is a stud or other unit of lumber that either receives the end of a header in floor framing, or that supports the free ends of floor joists, studs, or rafters. In wall framing, the trimmer is nailed against the final studs on either side of a door or window frame. At this point, the trimmers form the inside borders or frames for the door or window.

When the doors and windows are *rough-framed*—that is, when the openings have been laid out and the trimmers have been nailed in place—the top of the opening must be framed. This part of the framing is called a *header*. The header is usually composed of two lengths of 2 × 4s or 2 × 6s nailed together and installed over the opening. If the header is too thin, you can nail thin strips of wood between the larger pieces of lumber to get a perfect fit.

The *soleplate* is the 2 × 4 that is nailed on top of the subflooring. This 2 × 4 extends the length of the wall and rests on the box sill or joists. The studs are then nailed to the soleplate. (See Fig. 1-3.)

The *corner posts* are located at the corners of walls and partitions. These posts are made up of spacer blocks and three 2-×-4s nailed together. The three (or sometimes more) ordinary or common studs used in the corner post are designed to add strength to the corners. The corner posts are installed by nailing them to the subflooring, after which they are *plumbed* (located in a perfectly vertical position) and temporarily braced. (See Fig. 1-4.)

Where a partition meets an outside wall, you must nail in a stud that is wide enough to extend beyond the partition on both sides. This stud provides a solid nailing surface for the inside wall finish. (See Fig. 1-5.)

The *top plate* is the beam, usually a 2 × 4 that is nailed across the tops of the corner posts, to tie the studding together. It is also the link between the wall and the roof, and is often referred to as the *cap*. A second 2 × 4 nailed over the plate, doubling it, is called the *double-plate*. It serves to strengthen the top plate, and, thus, the entire wall.

Header joists are pieces of timber to which the ends of other joists are nailed to form a boxlike structure.

Cut-in and *let-in* bracing are often used in addition to the diagonal bracing. Let-in bracing is constructed by cutting out a 1-inch section of the edges of studs and nailing the 1-×-4 brace flush with the edges of the studs. Cut-in bracing is created by cutting a 2 × 4 diagonally into sections and toe-nailing each section between the studs. The finished product looks as if a single length of 2 × 4 had been nailed diagonally from corner post to soleplate.

Directions on construction of corner posts and T-posts are provided in Chapter 6.

PLANNING WALL LAYOUTS

You need to plan carefully to produce the most efficient and adaptable wall plans. The most obvious consideration is the size of the room or wall.

Many home owners have either bought or built a house that seemed to have adequately large rooms, only to find that, indeed, their furniture did not fit. In other cases, the number and arrangement of windows and doors created difficulties.

Fig. 1-3. If you have helpers, lay out the soleplate and top plate side by side, mark them for studs, and then lay out the rest of the wall frame, including headers.

For instance, a bedroom that is 18 × 20 feet would seem to be adequate for virtually any furniture. However, if this bedroom has three windows, three doors, and a fireplace, there will be room for little more than a queen-sized bed, a dresser, and a chest of drawers.

It is not unusual to have three windows in a single bedroom. If the bedroom is a corner room and if there is a fireplace, for proper balance there perhaps should be a window on each side of the fireplace. An additional window might be located on the other outside wall.

Fig. 1-4. The corner post is one of the most crucial elements of the wall frame. You can use solid posts or an assembled post, like the one shown.

Fig. 1-5. Use a partition junction post, like this one, wherever a partition wall meets an outside wall.

Although locating three doors in a bedroom might seem extreme, this would not be unusual if you have one entrance door, one closet door, and one door to the bathroom. You have a room that is chopped up considerably at this point.

The fireplace could be omitted, or course, but if the house is a two-story and if there is a fireplace in the den below the bedroom, it costs very little more to add a fireplace to the bedroom. Such an addition can be extremely welcome on cold winter nights.

In the case of a den, there might be one door to the outside, another to the hallway, and perhaps still another to the basement or to the adjacent room. There might be a cloak closet as well. As a result, there would be four doorways, plus window openings, and perhaps a fireplace.

If you are laying out the framing for an exterior wall, keep in mind that the orientation of the wall might be of considerable importance. For example, if you live in a rather cool climate and the wall is on the west side of the house, you might want a great deal of window space so you can take advantage of the heat of the afternoon sun. On the other hand, if you live in an area noted for its year-round warmth, excess window space on the west wall might heat the room and run up your cost of air conditioning. Many new houses are built without windows on the north side to reduce heating costs.

Consider aesthetics as well. If there is a spectacular view from one particular side of the house, you might want a large window on that wall.

Locate exterior doors at a sheltered area, such as a porch, to provide protection from wind and rain. Consider the height of the interior floor from the exterior ground level, and plan doorway locations to reduce the need for expensive stairways.

When you lay out windows in an exterior wall, consider the furniture to be used in the room, the purpose of the room, the heating system for the room, and the activities that are most likely to occur in the area outside the window, such as children's ball games, that could result in broken panes and similar damage.

Plan interior walls carefully. Ask yourself several questions before you commit yourself to the size and other plans for the wall. For what purpose is the room to be used? Will the room continue to be used for the same or a similar purpose? Will a bedroom later be converted to a game room, or will a den become a bedroom? If the room is for two or more children, how will the room be used when the children are old enough to each need a room of his own? If an older son or daughter will use the room, what happens when the child moves away to start his own family? Where will the thermostat be located in relation to windows, doors, and other draft-causing areas of the room? Will the wall plans allow for enlargement or partitioning? (For instance, if the family size increases and a playroom or den must be converted to a bedroom, can a closet or bedroom be added?)

In the case of exterior walls, if a room is intended for a bedroom, you will want windows placed high enough so you can locate the bed, for instance, on that wall. For interior walls, you will need to decide where the heat ducts will be; where plumbing, if any, will be connected; where electrical switches and sockets will need to be located; and whether a doorway could be added later if one is needed.

In planning both interior and exterior walls, you should think about the finished wall. Will you use sheet paneling on the inside, or will you put up gypsum board, diagonal sheathing, textured wall covering, or wallpaper? Will you use clapboard or bricks for the exterior wall?

Consider the standard size of finishing materials and incorporate these measurements into your wall's dimensions. You can thereby save the expense of wasted end pieces, and the time spent cutting and custom-fitting these supplies. For example, plasterboard comes in 4-foot widths. Planning the wall in 4-foot units simplifies your work by allowing you to use whole sheets. Because sheet paneling is tricky to cut and fit so that the seams are not noticeable, you should also use whole units of this material, whenever possible.

For exterior walls, if you are using clapboard, you can get more for your money if you plan the wall so that you don't have to cut off eight or ten inches from each board. If you are using bricks, plan the wall so that you can use whole bricks instead of halves or bats. The same is true for various types of shingles, shakes, or sidings.

You can also save money, time, and effort by planning your walls so that you can place studs at regular intervals and so that no two studs need to be close together. At the corners, for instance, if you must nail the next-to-last stud only a few inches from the corner, you will, in effect, have to waste a stud by nailing in another at the corner. You also will need to use an additional joist or rafter when it comes time to add a roof.

You can help yourself a great deal by planning so that a stud will be nailed where a socket or switch will be placed later, or by planning so that sockets will be centered under the windows. Carefully plan how the door will swing and how you can use that part of the wall most efficiently.

In summary, don't simply plan to have four walls, windows, and doors in a room. Visualize the contents of the room. Make a rough sketch of what furniture will be in the room and what the traffic flow will be with the furniture in place. If you plan to use a game room partially to hold a billiards or pool table, you must keep in mind that such a table is about 8 feet long. The cue sticks are 5 feet long, so you will need at least 5 feet of clearance on each end of the table. Therefore, the room must be a minimum of 19 feet long in order for players to use the table comfortably. If you are planning a home entertainment center, the room should afford ample seating for comfortable enjoyment of the center.

ECONOMY VS. FALSE ECONOMY

Most do-it-yourselfers have, at least in part, the idea of saving money as one of their motives. All too often, however, the apparent savings can be costly.

Prevent false economy in the wall-framing stages, before you are too far committed to change plans. One of the most serious forms of false economy is in the use of low-grade studding. Many lumberyards sell "economy-priced" studs for an attractively low price, but when they are delivered you might find that many have soft spots in them, the beginning of decay, knotholes, or similar defects. When you install these studs and then nail plasterboard to them, you might encounter problems.

In some cases, the nails will not hold because the wood is damaged, and you might have to drive in nails in several spots before you have a firm hold. You will also find that you need to do a great deal more patching and filling for the plasterboard, and that the "savings" has become costly.

If the end cut of the stud went through a knot, you will find it very difficult to nail the stud in place because the knot will fall from the wood when the nail is driven into it. The stud is thus virtually useless except for use as a cripple or for limited use in door framing.

Often the knot is in the middle of the stud, thus forcing you to use extra blocking to strengthen the stud. It is also difficult to nail switch boxes onto such weakened or defective studs.

If you decide to buy the cheaper studs, ask the lumberyard if you can pick out your own. Many of the discount houses will not allow selective purchasing, so you are better off buying the higher grades of lumber.

You can have similar problems in the purchase of other cheaper products. For instance, plasterboard might have been damaged when it was dropped or when some heavy object was dropped upon it, and the damage will not be visible until you unwrap the boards to use them. Many similar building materials are easily damaged by water, by prolonged exposure to damp weather or extreme heat, or by rapid fluctuation of temperatures.

Later, when you are finishing the wall and nailing up the paneling you bought at cut-rate prices, you might damage a sheet or run half a sheet short. When you return to the supply house you might find, to your dismay, that the paneling was sold at a good price because it was a discontinued item, and you cannot buy any more of it—anywhere.

When you start to erect the wall framing, carefully examine the subflooring for any signs of weakness, decay, or other deterioration. If you put up a wall on deficient subflooring, it will be only a matter of weeks before the soleplate is pushed slightly down into the subflooring by the weight of the wall. The wall will deviate from its square position by 1 inch or more, and even a fraction of an inch can make a difference when you are finishing the wall. Worse, after the wall is finished and the soleplate begins to sink, the molding in the room might start to pull away from its snug fit.

BEST MATERIALS TO USE

What is the best wood to use in wall framing? The answer can vary greatly, depending on where the wall is to be located. The wood used in an upstairs room partition does not deteriorate the way wood used in a basement can. Generally speaking, pine is good for any sort of wall framing that stays dry.

One major consideration is how much money you are willing to pay for excellent versus adequate building materials. Some woods that are marvelous for building are exorbitantly high in price, while many types of lumber that are easily affordable are not satisfactory for the job.

Basswood is one of the least expensive woods on the market. It is light, very easy to work, straight, knot-free, and true-grained. It does not warp or twist, and you can drive nails or screws into it very easily. On the other hand, it is very soft, it is not durable, and if installed before it is fully cured it shrinks greatly when it has totally dried.

Birch, a much better wood than basswood, is also more expensive. It is excellent for soleplates and top plates because it is very hard and durable, and does not bend or curve easily under pressure and strain, as long as it is not exposed to weather. (When wet, it decays fairly quickly.) It is a fine wood for interior work, however, and it is easily sawed and nailed.

Beech, although not quite as durable as birch, is a good wood for corner posts, soleplates, and top plates. However, it is more expensive than basswood or the pines and firs.

Douglas fir is one of the best and universally acceptable woods available. It is not terribly costly, and it is strong, straight-grained, and soft enough to saw and nail easily.

Its major defect is that it has a slight tendency to be brittle, especially when thoroughly cured.

Oak is one of the best woods for all types of construction, except for its high cost and its hardness. It is so hard that it presents difficulty in nailing and sawing. For corner posts and soleplates, however, it is a superior wood. Oak is also a very heavy wood, which you should consider before you choose to use it for the weight-bearing parts of a wall frame.

One wood that has received a poor reputation in building is poplar. This tree grows to great heights, is straight-grained, is soft enough to saw and nail easily, and is very light. It also holds nails very well. For studs and other uses in wall framing, it is not a wood to be overlooked. It does warp and bend rather easily when green, so make sure it is well cured before you use it. It rots almost immediately in constant contact with moisture, so do not use it close to the ground.

As a test, we put two wide poplar boards atop a masonry wall and left them exposed to heat, cold, rain, snow, and wind to see how long they would last. At the end of ten years, the boards were still sound and usable; however, if they had been enclosed in a wall and subjected to dampness and no sunlight, they would have lasted less than two or three years. One problem with poplar is that it will split rather easily, so you must nail with care.

Redwood is in great demand, partly because of its availability, but this wood is not as strong as yellow pine, and it is expensive. Its greatest advantages are that it is highly durable and it is resistant to splitting and shrinking. Redwood is also fairly soft and easy to saw and nail.

Yellow pine, as mentioned earlier, is one of the best woods you can buy for all-around construction use. Its heartwood is one of the most durable woods in use in American construction, but heartwood is at a premium. You will usually be able to buy sapwood from lumberyards, but this wood is, when cured, very hard, very strong, and exceptionally durable. It holds nails and screws well, and it does not warp or split to a great degree. Uncured yellow pine will warp under stress, but yellow pine lumber from a supply house likely will be cured adequately. We have heart lumber from yellow pine in our own 100-year-old house, and the lumber is still incredibly hard and sound.

Some woods have a reputation for toughness, hardness, and value in building, but often these woods are hard to work, so they are not well suited for building. Hickory, for instance, has long been admired for its hardness and toughness, but it is susceptible to insect attacks and decay when subjected to moisture.

Red and white cedar are very durable for corner posts but are too expensive for general use in framing. They also tend to be brittle and unworkable.

More important than the variety of lumber, however, is the condition of the lumber. You should feel free to return for exchange or refund any framing lumber that you find to be defective when it is delivered. As stated earlier, when possible you should oversee the selection of the studding before it is loaded for delivery. You should also refuse to accept any defective wood delivered to your work site.

CUTTING YOUR OWN LUMBER

Cutting our own wood with a chain saw provides the best lumber. We recommend this method highly if you have access to yellow pine trees or a similar variety of wood.

You will find that such trees are often available for the asking after fierce storms, since they are so top-heavy and shallow-rooted that they blow over easily. Many people will give you the trees if you will clear them off their property.

To chain-saw your own 2-×-4 or larger lumber, simply cut the trunk of the tree into appropriate lengths (8 feet for your studding, usually), and snap a chalk line down the center of the trunk from one end to the other. Using a chain saw with a very sharp chain, make a cut 1 inch deep along the chalk line. Then start all over and cut the log into two long halves.

Next snap another chalk line along the sawed edge of one half. Mark the line in far enough that you will have at least a 2-inch depth of good wood showing after the cut is made. Cut off the edge along the chalk line.

Now move over and make another chalk line 4 inches from the cut edge and saw off the section. You now have three sawed edges, and all you need to do is turn the unit of lumber on its side and mark the fourth edge. When you saw it off you have a fine timber that will probably be better than anything you can buy at the lumberyard.

A chain saw can be a very dangerous tool, though, and you should exert the greatest caution when using one. Never take any chances, and wear appropriate protective clothing. Such protection includes heavy gloves, goggles, ear plugs, and heavy boots. Observe all of the safety rules of chain-sawing at all times. The work can be hard and rather slow, but with practice you can saw a 2 × 4 or 4 × 4 in about five minutes, once the log has been cut and your cost will be less than $.10 per unit of lumber.

INSPECTION AUTHORITIES AND BUILDING CODES

Inspection regulations and building codes vary so much from state to state that we can only comment on general topics. You should check with your local authorities before you start to build a house, add a room, or modify the outside of your residence in any way. In some states, you cannot change the outline or profile of your house without a building permit, and in most cities a permit is required before you can erect a new building inside the city limits. Some counties have similar regulations.

Most building permits are inexpensive. Their cost is generally based on the cost of the building or addition.

If you do start a building without a permit, you might be told to take down the structure, secure a proper permit, and notify the building inspector so that he can examine the footings, the foundation, and other key points of the building.

Electrical wiring and plumbing are almost universally subject to municipal or county regulations. Do not try to circumvent these rules and regulations. Faulty plumbing installation can result in severe damage, great problems, and chronic frustration. Faulty electrical wiring might prove to be fatal or injurious, and your house could be damaged or destroyed.

Do not assume that inspection authorities are needless members of a bureaucratic government. These officials provide a very necessary service. If your structure is defective, it is much better to learn about it and correct it than to suffer the consequences.

You can help yourself by reading your local building codes and familiarizing yourself with the specific requirements for the type of construction you are planning. Do so before you start to shovel, saw, or nail, and certainly before you make arrangements to purchase materials or employ carpenters or helpers.

In some states, you can use lumber from another building in home remodeling or construction. You can buy salvaged lumber readily and use it in your projects, and you can save a considerable amount of money by doing so. (See *Building with Salvaged Lumber*, TAB No. 1597, for detailed information on locating and using salvaged lumber.)

In many states, however, you cannot use salvaged lumber in a load-bearing wall. If you plan to use such lumber, check with the local authorities to determine whether any such restrictions are in effect in your area. Generally a telephone call or visit to a local library or courthouse will provide the needed information.

When you start construction, after securing a building permit and having footings inspected, don't hesitate to call the local authorities on any building matter you have questions about. You might not know, for instance, that in many states any room that has a closet might be classified, according to the building codes of the states in question, as a bedroom. Minimum capacity of a septic tank, in many states, is predicated upon the number of bedrooms in a house. So keep in mind before you lay out a wall or room that includes a closet that you might be asked to update septic tank size, since you have technically added another bedroom.

To be safe, check with local inspectors. It would be highly inconvenient and costly in time, money, and effort to have to dig up your existing septic tank and replace it with a larger one if you only intended to add a den or game room to your house.

The logic behind the closet-bedroom policy is that you might increase the size of your family by having relatives move in or by adding children, and the present septic tank would not be able to accommodate the family's needs.

Another argument from the building inspector's point of view is that you might have a nominal two-bedroom family, but if you should sell your house to a larger family, the game room or den (with the closet you added) could be converted to a bedroom. The septic tank capacity would not be adequate for the larger family.

The same sort of argument applies if you decide to add another bathroom to your house. You might find yourself facing the problem of unexpected expenditures of time, money, and energy.

Even if you are planning to add only a family room—no closet or extra bedroom—the inspector's office might determine that you do not have adequate electrical fuse-box capacity for more wiring and might ask you to upgrade your electrical system.

This discussion is not intended to discourage you from starting that special project. On the contrary, it is only intended to encourage you to go through all the proper channels before you spend money and start to work.

Believe it or not, the building inspector is one of the best friends you have in home building or remodeling. He has the background, insight, knowledge, and duty to assist you in your building projects. His primary duty is to assure you and your family that you are not exposed to unnecessary dangers because of your lack of information concerning building safety regulations. Welcome the inspector and his suggestions: they are for your own benefit and welfare. Do not argue with the inspector when he tells you what must be done. By law he is required to enforce the regulations.

You actually can save money in your real estate taxes if you secure the necessary building permits because in many areas the extra taxes levied against your real property are calculated on the cost of the extra room you are adding to your house. If you are building the room yourself, the cost will be very low, and the lower the cost, the lower your tax base on the addition will be.

BUYING MATERIALS

You also might be able to save money if you can find a local supply house that will give you contractor's prices on the materials you buy. You can argue that the room you are adding will result in large enough sales to justify the cost reduction. You might get a special price on every item you buy, and if you are adding a complete room, the money you save can be a rather significant amount. (Do not ask for a contractor's price if you have not yet secured your building permit, however.) If your purchases amount to a large sum of money, delivery will likely be free.

Before you buy materials, calculate what you will need. For paneling or plywood purchases, measure the room and count the number of full panels needed. Add another panel for every two windows you need to panel above and beneath, and still another for every six doors to panel above.

When buying studding, figure four studs for each corner and another for every 16 inches of wall space. Add three more for each partition post or wall-junction post. Measure the length and width of the room and multiply the total number of linear feet by three for top caps, top plates, and soleplates.

Add 20 feet of 1-×-4 lumber per corner if you plan to use diagonal bracing. Add 16 feet of the same lumber for each door, and 18 feet for each window.

To determine the amount of insulation, count the spaces between studs and multiply the total by 8 feet. You will need to buy 8 feet of 2-×-6 or 2-×-8 lumber for every window and door in your building project. For each room, add another dozen 8-foot-long 2 × 4s to use as blocking, trimmers, and sills.

Finally, you can buy nails by the pound, figuring three pounds for each kind of nail you will be using regularly.

2
Necessary and Optional Equipment

ALL BUILDING PROJECTS REQUIRE TOOLS OF SOME KIND. IF YOU DO
not own a modest selection of tools already, you should plan to buy or rent the ones you
will need for your wall-framing job. Some tools are absolutely necessary whereas others
are desirable or optional.

BUYING VS. RENTING TOOLS

The number of tools you will need depends in part on your health, energy, strength,
time and money. You must decide whether to rent or buy tools in light of your own particular
circumstances.

If you have plenty of time and if you are in good physical shape, you might not mind
hand-sawing 2 × 4s and larger timbers. If you are in a rush to get the job completed
or if you are in less than good shape physically, or if your spare time is severely limited,
it might be to your advantage to buy some power equipment.

Almost every town or hamlet of any size has at least one tool rental business, and
you might find that you can rent tools much less expensively than you can buy them. Before
you make a choice, however, consider several questions: are the tools you plan to rent
equipment that you will need only one time, or will you have occasion to use them regularly
in future handyman jobs around the house? What is the difference between the cost of
renting tools for two weeks and the cost of buying the tools? If you rent tools, are you
responsible for reasonable damage to them? If you rent tools, how many of them will
lie unused for days at a time—days in which you still are paying as much as if you were
using the equipment hourly? Can you rent only one or two items at a time?

If you are planning to buy certain tools, you should consider whether you will use
the purchased tools often enough to make the purchase practical. Nearly everyone has
an occasional need for a hammer, screwdriver, and adjustable wrench, but how many

people have a regular need for an auger, circular saw, band saw, or hydraulic jack? If you find that a chain saw will be very useful to you on the project, will you need it often enough later to saw firewood or clear away fallen trees to justify paying for the saw?

A nail gun is a case in point. One carpenter, using a good nail gun, can nail an entire wall frame assembly while another carpenter working with a simple claw hammer can drive only half a dozen nails. The nail gun is fairly expensive, however (about $250 for one of good quality), and an air compressor (another $250) is necessary to power the nail gun. You will have to decide whether the time and labor saved can justify adding $500 to the cost of the house or room.

Every do-it-yourselfer must answer these and other questions on the basis of personal needs and circumstances. You might make the decision based only on your current monetary conditions.

NECESSARY TOOLS

There are several tools that you will need regularly during your wall framing project. When you buy these items, be sure to purchase quality tools. If you choose to buy a hammer, get a good one. You can buy a cheap hammer, but the handle might work loose quickly or you might break the handle if you use the hammer to pull out nails.

The same principle holds true for almost any equipment you will use during your wall-framing projects. If you buy second-rate tools, you will probably have to replace them before you get your money's worth of use from them. If you have to buy tools twice, you could have bought good tools as inexpensively in the first place.

Many recognized brand-name tools carry a refund policy. The manufacturer will either return your money or replace the tools if they prove to be defective. A tool that breaks during proper use is indeed defective. Do not hesitate to take the defective item back and ask for an exchange or refund.

If you are paying assistants or co-workers, don't attempt to economize by not buying the following items. If you are paying a helper $8 an hour yet refuse to buy a $25 tool, bear in mind that you will pay far more than the cost of the tool in labor expenses for the helper, who will produce less work with substandard equipment.

One indispensable tool is a good *claw hammer*, that is, a hammer whose head has one forked end. If you are working alone, one hammer will generally suffice. If you have a helper or two, it is cheaper to buy extra hammers than it is to pay someone to wait while you finish with the single hammer and pass it to him. Check to determine that the handle is sturdy and strong, or buy a hammer with a metal handle.

Even if you are working alone you may find a second hammer is desirable, particularly if you don't have at least two sizes of crowbars or wrecking bars. In pulling out nails that have been driven deep and embedded into the wood, a second hammer can be a wonderful investment. If the nail is so deep that neither the claw hammer nor a wrecking bar can grasp the nailhead, you can hold the claws of one hammer in position and strike its head with a second hammer, forcing the claws far enough into the wood to grasp the head of the nail and allow you to pull it out. (See Fig. 2-1.)

If a nail is stubbornly embedded and you have trouble extracting it, you can use the head of one hammer, laid flat, as a leverage fulcrum to help you remove the nail.

In addition to the hammer or hammers, you will need a good *square*. Get one that is at least 2 feet long on the longer side and a foot and a half on the short side, and be sure that all of the numbers and gradation marks can be seen clearly. It is very inconvenient and time-consuming to have to peer at fractional inch marks while you are holding a timber in place. It is even more inconvenient and expensive to waste a length of lumber by cutting it wrong because it was marked wrong. Nailing together pieces that fit poorly results in unsightly carpentry work. A smaller square, one that is 6 inches long, is handy for work in small spaces.

A *sliding T-bevel square* is amazingly useful when you need to set or try angles other than right angles. You can set almost any angle with this tool and save time and lumber.

Fig. 2-1. To remove nails with sunken heads, use one hammer to drive the claws of a second hammer into the wood slightly and over the nail head. You can then extract the nail.

You obviously need at least one good saw, perhaps more. To save money, you can get by with one good *handsaw*, one that makes a smooth and fine cut. A handsaw cuts more slowly but doesn't produce rough edges and splinters.

If you can justify the expenditure, add a *circular saw* to your list of tools. These saws are inexpensive and are very fast and efficient. If you buy only one blade, buy a multi-use blade that rips (with the grain) or cross-cuts. If you wish, you can purchase blades for a variety of purposes. A blade that will cut metal, masonry, stone, and other hard substances is available if you need to shape any of these materials for your framing work. (See Figs. 2-2 and 3.)

At one point you will need a *pry bar* or *crow bar*, sometimes called a *wrecking bar* or *construction bar*. You need one that is at least two and one-half feet long. These bars have a multitude of uses, ranging from extracting nails to prying or levering timbers into place easily and accurately.

A *level* is indispensable. You will need it to set perfectly vertical corner posts and level top plates. Buy a level that has a clear, easy-to-read viewer and a bubble that offers enough contrast so that it can be read easily in shade and other poorly lighted areas. Levels of several lengths are available. The three-foot length is ideal, though a four-foot level is more accurate. The disadvantage of the longer level is that it is often difficult to maneuver into tight working spaces. If you need to get a reading over a wide expanse, as from one wall frame to another, you can find a straight and evenly cut 2 × 4, place it across the expanse, and set the level atop the timber.

A good ruler or *tape measure* is also invaluable. You will need to mark and cut exact lengths of lumber, so buy at least a 16-foot tape, with clearly marked inch and fractional inch locations. The tape should have a metal tip that can be hooked over the end of a board for easier measuring. A tape that can be locked in place and unlocked with pressure from your thumb is especially convenient. Another nice feature of many tape measures is the clip that can be used to fasten the tape to your belt or shirt pocket for ready access.

A good pocketknife is a necessity. In any kind of outdoor work it is only a question of when, not if, a pocketknife will come in handy. You need such a knife far more often than you would believe. It need not be an elaborate or expensive one, but you will find dozens of uses for a good knife.

Essential safety equipment includes a pair of safety glasses. These glasses protect your eyes from flying particles while you work. You need work only a few minutes without eye protection to see how often foreign and potentially damaging particles hit in or near your eyes while you handle lumber, hammer, and saw. For example, a circular saw kicks bits of sawdust toward your face. This is very dangerous to unprotected eyes. Even if you are using a handsaw, slight gusts of wind can blow sawdust into your eyes. When you are nailing in the loose ends of cross-bridging, lying on your back, the force of the hammer might cause debris to fall into your face. The same problem might occur as you stand and reach upward to nail in the top ends of studding, if you are erecting a wall frame in the piece-by-piece fashion.

Wearing goggles is admittedly annoying when the heat causes your body heat to fog up the lenses. It is far easier, however, to stop to wipe the glasses than it is to experience a serious eye injury.

Fig. 2-2. If you need to square edges of stones for use in foundation walls or piers, clamp the stone to a stable surface and then use a circular saw with a stone-cutting blade.

Fig. 2-3. After you chalk or mark a straight line across the stone, you can saw a straight line easily with the special blade.

Wear good gloves when you are handling timbers, paneling, and plywood. Splinters from the edges of these building items can embed themselves into your flesh or tear your skin.

OPTIONAL TOOLS

The tools listed above are the barest of necessities. They will enable you to frame a wall; however, you will find the work to be much faster and easier if you add a few more items.

First, a smaller *crowbar* is useful. There will be many times when you need one, when the longer one cannot be used. You can save hammer handles and your energy by using a small crow bar.

If you bend a nail while toe-nailing a stud, you will find that you cannot get the full-sized crowbar or a hammer into the space. Thus, one of the foot-long crowbars can be useful. The larger crowbars, or wrecking bars, are very useful for pulling large nails.

You can also make excellent use of a *plumb bob*. This device, a shaped weight attached to a long cord, is perfect for making certain that a wall is perfectly perpendicular. It is also good for marking center points on floors, soleplates, and other areas of work.

A *fold-up rule* is also handy for measuring those hard-to-reach spots, so you do not need to climb a ladder to secure an exact reading. These rulers can be folded into a small size and carried in your pocket easily. Unlike the tape measure, the ruler is stiff enough to reach a considerable height without bending and falling. You can fold it at any angle you wish and then extend it to measure the point in question.

You also can use it to mark boards or braces. If you have difficulty cutting a piece of lumber at the exact angle you need, you can bend the ruler to conform to the inner or outer side of the angle and, leaving the ruler exactly as it was bent, mark the lumber and saw accurately.

A pair of *ladders* is highly desirable. Two useful ladders are a straight-rung ladder 8 to 10 feet long and a freestanding eight-foot stepladder.

Another handy item is a medium-sized *chain saw*. Although it does not produce a smooth cut, a chain saw is fast, and inexpensive to run. It is especially useful if you need to cut through heavy timbers quickly, or if you are working without access to a source of electrical power. (See Fig. 2-4.)

If you plan to bolt joists or sills and plates together, you may need an *auger* and a *keyhole saw*. The auger, or *brace-and-bit*, will enable you to bore the exact size of holes you need for the bolts. If you need a larger hole than your bit provides, you can use the keyhole saw to enlarge the hole.

An *electrical drill* is another tool that has a wide range of uses. Its major use is for boring holes into wood, but you can also get bits for boring through metals and masonry. This latter use is particularly helpful when you need to bolt sills to foundations. (See Fig. 2-5.)

You can also buy *sanding wheels*, *emery wheels*, and *sanding discs*, all of which have their uses. The emery wheel can be used to sharpen blades of other tools, or to cut away edges of studs or other timbers that are uneven, rough, or a fraction of an inch too wide. If you accidentally make a rough cut with the power saw, you can smooth the splintered edge with one of the sanding attachments in a matter of seconds with virtually no effort. (See Fig. 2-6.)

Fig. 2-4. Use a chain saw for timbers too thick to cut with a circular saw. If the chain is truly sharp, you can get a very straight and smooth cut.

Fig. 2-5. You can buy a medium-speed drill and several attachments for boring large holes for wires or pipes inside the wall frame.

Fig. 2-6. Another attachment for a drill is the sanding wheel. You can smooth a length of wood easily and rapidly by using the special disc with sandpaper pressed on it.

Other useful tools include a *plane*, *gouge*, *chisel*, and *punch*. The plane can trim down a wood surface for a better fit, or smooth it generally. A gouge is often helpful when you need to round out a hole. A chisel functions similarly but it can also make small emergency cuts. The punch sinks nail heads so far that they will not interfere with later nailing along that same surface. (See Fig. 2-7.)

Sawhorses are helpful items in any kind of carpentry that requires cutting. They are especially useful for the final wall finishing work. You can buy sawhorses or you can make your own in a short time, with little cost.

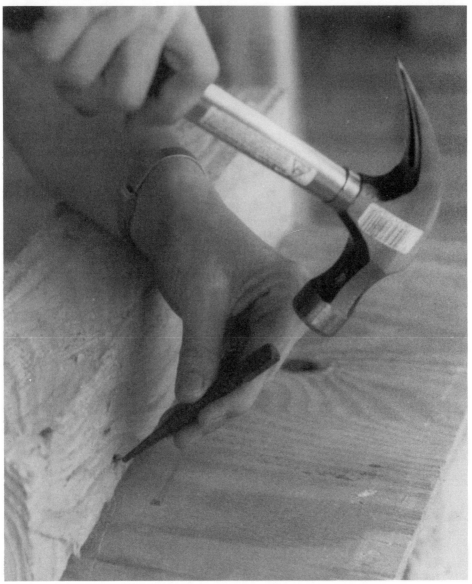

Fig. 2-7. Use a punch to sink nail heads slightly below the wood surface.

You will need a pair of sawhorses, typically each 5 feet long. To make each sawhorse, start with a length of 2 × 4 5 feet long. Cut four more pieces of equal length for the legs. Vary the length to suit your height and to provide the most comfortable level for a work surface.

To cut the legs at the correct angle, lay the 5-foot length of 2 × 4 on edge. Hold a shorter length of 2 × 4 so one end is against the very end of the first piece and so the other end of the 2 × 4 you are holding is at a point 15 inches from the other piece of lumber. Mark the piece you are holding by running a pencil along the end edge of the piece on the floor or ground. Saw along the mark and your angle will be perfect for the leg of the sawhorse. (See Fig. 2-8.)

When you have sawed the angle, mark the other three legs by the first cut. Turn the 2 × 4 on the ground or floor on its side and nail the first leg to it an inch from the end. Now nail the other leg on that side at the other end, then turn it and nail on the other two legs. (See Fig. 2-9.)

Stand the sawhorse on its feet and position a short piece of 1-×-4 board so that the top of it is against the bottom of the 5-foot 2 × 4. Mark it along the outer edges of the legs. Saw the board at the marks, then nail it where you held it for the marking. Do the same at the other end, and the first sawhorse is ready for use. If you wish to strengthen it even more, you can mark and saw brace pieces. Nail them halfway down the legs. (See Figs. 2-10 and 2-11.) Repeat the process for your second sawhorse.

The sawhorses are ideal for many kinds of work. They elevate the lumber to be sawed or nailed so that you can work in a standing position instead of in awkward, cramped, and tiring positions. (See Fig. 2-12.) When it is time to put up the wall covering, you will find that the sawhorses are perfect for holding plasterboard or paneling.

If you are working alone, a 6-inch *C-clamp* will help you a great deal. Instead of trying to hold a long timber and nail it at the same time, you can clamp the timber in place and nail it with ease. If you are sawing, you can clamp the wood in position and devote your full attention to the saw blade. (See Fig. 2-13.)

It is always useful to have a good *screwdriver* and good pair of *pliers* on hand, although these are not essential.

If you want your square corners to be really smooth, you may wish to invest in a *triangular file*. This file has three flat sides. It is excellent for clearing out any corners that may have splinters or rough edges showing. Other types of files are good for sharpening blades of circular saws and for sharpening chains for gasoline saws.

You can buy a *combination square* set, surely one of the best tools available on the market today. This device is actually several tools in one. It can be used for squaring, laying out angles, drawing parallel lines, measuring depths of cuts or slots, or locating exact centers. The combination square usually has a square head, protractor head, a center head, a scriber for marking cuts, and a level built onto the device. (See Fig. 2-14.)

NAILS AND LUMBER

Once you have obtained the required tools and as many of the optional ones as you think you will need, all that remains before you go to work is the task of getting the nails and lumber. You typically will need a quantity of nails in sizes from 8-penny to 20-penny. Nails are available from the 2-penny size to the 60-penny size.

Fig. 2-8. Hold the leg of the sawhorse in position and mark the saw line by using a felt-point pen, or a pencil. Then saw along the marked line.

If you are confused about nail sizes, remember that in many instances nails are labeled as, for example, 16d, which means 16-penny. The "penny" designation refers to the number of nails per pound. The larger the nail size, the fewer you get in a pound and the thicker the nail shank becomes. Your major concern, however, is the thickness and holding power of the nails you will put into your framing.

Nail lengths are as follows: 2d, 1 inch; 3d, 1¼ inches; 4d, 1½ inches; 5d, 1¾ inches; 6d, 2 inches; 7d, 2¼ inches; 8d, 2½ inches; 9d, 2¾ inches; 10d, 3 inches; 12d, 3¼

Fig. 2-9. Nail the leg flush with the edge of the sawhorse spine, or indent an inch and nail support pieces across the legs.

inches; 16d, 3½ inches; 20d, 4 inches; 30d, 4½ inches, 40d, 5 inches; 50d, 5½ inches; 60d, 6 inches.

You would rarely use nails larger than 20d or smaller than 6d. In fact, you will use 6d nails only to tack up a temporary strip of wood as a marker or for a similar purpose.

Fig. 2-10. When the legs are attached, nail braces to hold the sawhorse stable while you work.

Fig. 2-11. This sawhorse was made of lumber cut with a chain saw. Total working time was about half an hour, and total cost was less than 25 cents.

You will seldom need or use nails larger than the 20d lengths, which are used to nail down soleplates. A few larger nails might come in handy if you are going to nail joists or sills as preparation for your wall-framing work.

Nails with a smooth shank do not have quite the holding power as nails with a grooved or spiraled shank; however, you may buy whichever type you prefer.

Many nails are actually lengths of wire topped by a head for ease of driving. Some of the larger spikes are hardened metals, as are cut nails and other specially prepared ones. The nail gets its holding power when it is forced into the wood, where it displaces the wood fibers from their natural position. The structure of the wood causes the fibers to push back as they attempt to regain their former position. It is this pressure on the shank of the nail that causes it to remain in place. The harder the wood, the greater the pressure of the wood fibers.

Nails or wood fibers that result in less pressure therefore have less holding power. In soft wood, for instance, the fiber is looser than in hard wood. Less pressure is created and, subsequently, less holding power.

For these reasons it is imperative that you avoid using any wood that has suffered decay. Decay is the destruction of the wood fiber, and without the fiber the wood cannot exert any pressure at all against the nail shank.

The thinner the shank of the nail, the less fiber will be displaced and the less pressure will be exerted. Therefore, do not choose to use nails that are made with a very thin shank.

Fig. 2-12. These sawhorses, which were made from dressed lumber at a cost of about $12, can be used to hold lumber for sawing and for a multitude of other building purposes.

Once you have bought your nails, you have completed your list of equipment. Before you begin to work, it is wise for you to gain some familiarity with the various types of walls and the advantages and disadvantages of each type.

In Chapter 3 we discuss several wall types and explain why some are clearly superior to others. Some of the types of walls discussed are not at all efficient. They are included, however, to provide information pertinent to all sorts of wall framing.

Fig. 2-13. One of the handiest and most inexpensive tools you can buy is a large C-clamp. The clamp should be at least 6 inches wide.

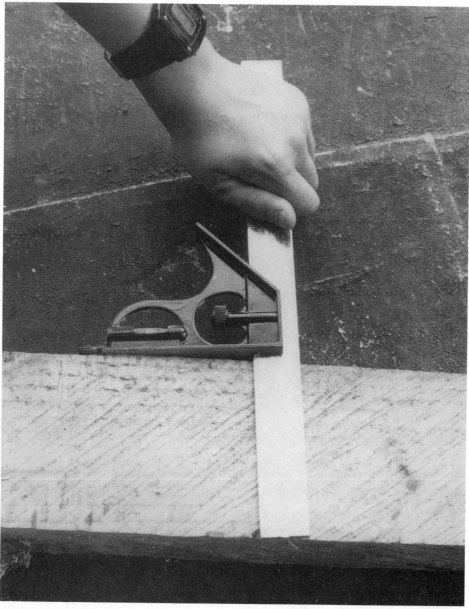

Fig. 2-14. A combination square, which is a multipurpose tool, can be bought for about $4.

3

Kinds of Walls

IT IS AN UNDERSTATEMENT TO SAY THAT THERE ARE SEVERAL KINDS of walls. There are, in fact, dozens of types of walls. Because carpenters commonly construct only a few types, however, we will discuss only these most popular walls in this chapter.

Some walls do not require actual framing, but they do require either levels of skill and knowledge, or materials, that are not commonly available to the typical do-it-yourselfer. In many instances these walls are not particularly well suited for various parts of the country or for residential dwellings. Some walls mentioned here have only historic and cultural value.

Many walls have labels or names that vary from one part of the country to another. Variant names and labels are used when pertinent.

BEARING AND NONBEARING WALLS

The two most important kinds of walls have been discussed in an earlier chapter: bearing and nonbearing walls. While the nonbearing wall serves primarily as a partition, bearing walls are crucial to the strength of the house. We have already noted that building codes vary so widely that it is advisable for you to check with local authorities for the latest changes.

In some areas, for instance, previously used lumber cannot be used in a bearing wall. The reason for this ruling is that some of the salvaged lumber available has been weakened by dry rot. According to these building codes, however, an incredibly sound and strong 4 × 4 of heart pine that has been used for less than a year in a new house may not be used as a stud in a bearing wall in a new room for your house. At the same time, though, the same region might have a building code that permits the carpenter to install studding 30 inches or more apart.

Whereas such rulings are difficult to understand, they are still enforceable. You should make every effort to stay within the code of your area.

Remember that a bearing wall must be strong enough to support the ends of rafters or joists that in turn carry the weight of the roof or the rooms above it. Bearing walls must also be strong enough to withstand high winds and any other force that is commonly exerted against them. The bearing wall must, therefore, be thick enough to guarantee stability.

Most home builders agree that the bearing wall in homes more than three stories high would have to be so thick as to be highly impractical unless special reinforcement were used. In some cases a steel reinforcement is used, and lintels made of angle steel are installed over windows and doors and over any arch or wide opening anywhere in the house. This lintel construction is widely used in brick or other types of masonry homes.

UNCOMMON TYPES OF WALLS

There are many types of walls that are either uncommon or are unsuited in residences, or are difficult to construct. Among these types are the adobe wall, the rammed-earth wall, the cob wall, and various other types of earth or soil walls. These do not require any framing except for forms that are used to hold the material in place until it settles and hardens sufficiently for the wall to hold its shape.

Adobe walls are made from sun-dried bricks which are composed of a clay soil and some type of fibrous material such as straw. The straw adds no strength to the brick; instead, it helps to prevent shrinkage during the curing process.

The major advantages of adobe bricks in walls is that the clay offers good insulation against both heat and cold, and where there is a scarcity of trees or other building materials the adobe material is generally available in abundance. Adobe construction is also extremely inexpensive. The bricks are easy to make and the process is fairly rapid. (See Fig. 3-1.)

To make adobe bricks the modern way, wet a large amount of suitable soil (with clay content rather than organic matter, roots, etc.) and let it stand a day or two so that all clods break up. Then work in the straw or other fibrous matter with a hoe, much as you would mix concrete or mortar. The traditional way to work the mixture is to then trample it with bare feet.

When you obtain the proper consistency, form the mixture into bricks. Dry them in the sun for about two weeks. Lay them flat at first. After a week turn them on edge to ensure uniform drying.

The major disadvantage of adobe walls is that unless the watershed is directed away from the house, the lower courses of brick will disintegrate in the excess moisture. If the bricks are not laid on a stone or concrete foundation, the ground water will infiltrate the lower courses, causing them to disintegrate.

The old-fashioned *cob* wall is similar to the adobe wall. This type of wall was used in ancient England and is still used in a rather limited fashion. The word itself means a lump. The wall is built by using stiff mud which is piled and shaped into a wall without any form of skeletal framing. The advantage is that the cob wall is very inexpensive, but the numerous disadvantages include the disintegration process.

The mud and concrete wall is made by building a frame, or form, into which very wet clay mud is poured. As soon as the wall is dry, the forms can be removed.

The *cajob* wall is similar, except that the frame is actually on the outside of the wall and is left in place. A frame is built and put into position. The hollow frame then is filled with damp earth which is packed and allowed to dry. The frame, which can be constructed of either wood or concrete, is left in place and the filler acts as insulation.

Fig. 3-1. Mix the water and red dirt or clay thoroughly, then add pine needles or similar fibrous materials. Mix the pine needles into the mud fully, then shape the bricks and lay them in a sunny, dry place to harden.

Hollow or *cavity* walls, as their label indicates, are built with a hollow or dead air space inside. These walls are usually composed of concrete blocks, bricks, or poured concrete, and the air space acts as a barrier against heat or cold. This type of wall is also called a *core* wall. (See Fig. 3-2.)

A *faced* wall is made of masonry materials, often of cement or construction blocks, with another material bonded to the wall so that the two are inseparable. The *veneered* wall, on the other hand, is similar, but the backing and facing are simply attached rather than bonded. The facing materials range in thickness from two inches or less.

A *curtain* wall is any exterior wall that does not serve as a bearing wall. It is used as a decorative element and as a barrier against wind and rain. Such walls consist of basic framing covered by thin-gauge metal or wood.

Sometimes the curtain wall comprises a simple 2-×-4 frame, with studding, sole and top plates, and bracing sufficient to meet building codes. Weatherproofing paper, then clapboard or other exterior wall covering, completes this wall. (See Fig. 3-3.)

The major advantages of these uncommon walls are that they can be constructed quickly and inexpensively, and of materials that can be obtained easily. Other types of walls in a house, though, must be seen from a deeper perspective.

BASEMENT WALLS

Exterior basement walls, or perimeter walls, are usually built of poured concrete or concrete block. Interior basement walls or partitions, however, are often wooden walls with regular studding.

Build the footings of exterior walls wide and deep enough to meet or surpass local building codes. The walls should be at least six inches thick and should have drain tiles and gravel beds outside them to divert any water seepage, even if the landscaping provides drain measures.

Treat the wall surfaces with some type of water barrier, such as an asphalt mixture or a coal tar product that will keep water from seeping through the blocks or concrete. Before you begin the coating clean the walls thoroughly of all dirt or mud and other foreign materials. Apply the covering generously to the mortar joints.

Lay gravel covered by sheets of plastic under the concrete floors to prevent water seepage. When possible, allow for adequate venting of all basement walls.

If you do not take precautions against water seepage and water vapors, and you use basement rooms as bedroom or den space, you will find that mildew becomes an immediate problem and your bedclothes and other fabrics will stay uncomfortably damp much of the time. If the area is used for storage, mildew is again a problem. Any papers or cardboard boxes will deteriorate very quickly, and everything will pick up an unpleasant musty odor.

Dampness in a basement also damages paint, paneling, plasterboard, and all types of timbers. Dampness attracts roaches, water crickets, and other insects that damage walls and other property in the basement. Greatest damage, however, can occur inside the walls where decay can spread unseen.

Wood set against concrete will decay quickly. If the concrete is damp, the decay starts even faster and progresses more rapidly. The cause of the decay, even if it is not visible, is moisture, which is absorbed by all kinds of materials from rocks to wood.

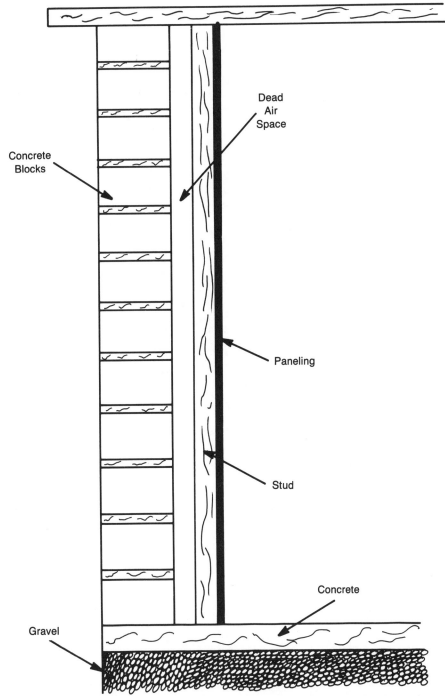

Fig. 3-2. To construct a hollow or cavity wall, start with the regular footings, then pour a concrete slab. Build the masonry wall atop the slab and then erect the wall frame 3 or 4 inches from the masonry wall.

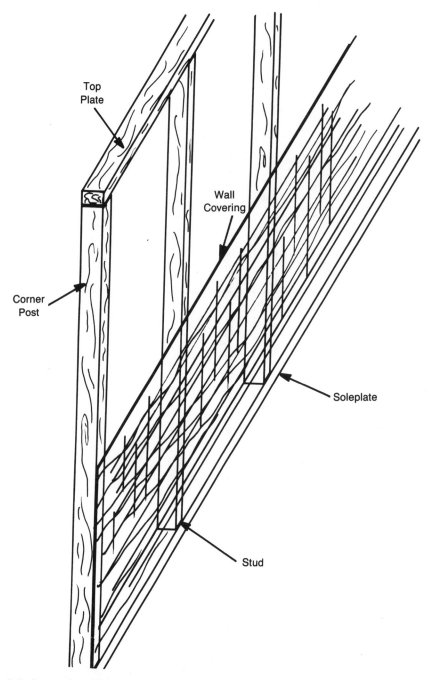

Top
Plate

Corner
Post

Wall
Covering

Soleplate

Stud

Fig. 3-3. A curtain wall is essentially a partition or nonbearing wall. The diagram shows the manner of construction.

If plasterboard absorbs dampness, it becomes soft and will not exert pressure against nail shanks. The plasterboard, therefore, will start to sag and pull loose from nails. Eventually it will fall. This is especially true of plasterboard used on ceilings. Moisture is the enemy of all types of plasterboard, however. The boards paper covering and any paint or wallpaper will deteriorate, and the entire wall might be ruined within a few weeks.

Paneling, too, collapses quickly if it is exposed to too much dampness. It will buckle and warp, and then pull loose from nails and other fasteners.

Inside the wall the studs will start to decay at the bottom, and soon they will start to sag. You might notice that one corner tends to drop slightly, or you might feel a thin coating of water on the walls. If you can wipe your hand across a wall and see a thin watery smear, you will start to see more serious signs of water damage within weeks.

To prevent part of the water problem, in addition to using gravel, drain tiles, and water barrier treatment, spread thin-gauge metal, water barrier paper, or plastic under all wood that comes in contact with the floors or with damp concrete or masonry walls.

STRENGTHENING BEARING WALLS

We have stressed the importance of bearing walls throughout this book for excellent reasons. Your house is no stronger than the bearing walls, which support not only their own weight, but the weight of the rest of the building above them. Bearing walls can be strengthened however, by taking a few inexpensive, quick, and easy precautions. These basic steps provide strength and durability that are well worth the effort required.

One of the most basic precautions is to examine the ends of all of the studding. If the end is defective, do not use the stud. You can cut off the defective end and use shorter lengths for cripple studs or for braces. If you feel that you absolutely must use a stud that seems fragile, put the weak end at the top rather than at the bottom of the wall so it will be located away from moisture.

Examine the 2 × 4s that will be used for soleplates, too. If the timber appears to be unsound, use it for other purposes. Any 2 × 4s that have cracks, abnormally dark areas, or soft spots will not be suitable for soleplates.

Do not use weakened timbers for corner posts, either. These members hold the corners of the wall firmly in position and keep them square. Do not use any beams that look as if they will sag or in other ways weaken the corners.

If you cut a defective stud and later use parts of it as one of the cripple studs, do not use the weakened portion. Remember that your windows and doors represent the largest openings, and therefore the greatest unsupported expanses in the wall. Using a defective cripple under a window is an invitation to problems.

Your wood should be properly cured, and you can tell very easily if it has been kiln-dried properly. If you are using pine studding, you will find that the wood is very light. So if you pick up studs that seem unusually heavy, the odds are great that the wood has not been dried sufficiently.

Another way of detecting green wood is by looking for the presence of unusually resinous surfaces. You will know resinous wood simply by touching it: it will be very sticky, and the gumlike substance will adhere like glue to skin, clothing, and other surfaces.

Some resin will be present in nearly any pine wood, but the amounts will be small and generally dark-colored. If the resin is sappy or thin, it is likely that the wood needs to be cured more. Wood that has been on the side of a house for decades, however, at times will still have resin in it.

Uncured wood has several disadvantages. First, it will dry in its installed position and shrink considerably, leaving fairly large gaps where there once was a tight fit. Remember that nails hold wood securely because the fibers push against the nail and wedge it tightly in place. If the wood shrinks, however, the fibers that surround the nail will shrink too. The result will be nails that fit loosely in the wood, and the secure nailing that held the green wood in place will be lost.

Green wood also bends and warps easily, especially if it is installed under stress. If there is weight on a green top plate, for example, the plate will bow downward and cause the ends to rise slightly, pulling the nails from the wood to which the plate was nailed. Green wood above and below windows and above doors will also bend, bow, and sag, and will ruin doorways and window fits, and often it is necessary to rehang and reframe windows and doors.

One of the best ways to strengthen corners is to use diagonal bracing. Chapter 6 contains suggestions and illustrations for the use of corner diagonal bracing.

One of the best ways to strengthen a corner or any part of the wall that has a great deal of stress upon it is by using a sheet of plywood. The insulation properties of plywood are not excellent, but its strength is great. If you choose to do so, you can use a 4-foot sheet of plywood on both outside surfaces of a corner and the corner will be very tight, strong, and stable. (See Fig. 3-4.)

Steel diagonal bracing is another fine way to strengthen corners of bearing walls. Fasten the steel bracing strips to the top of the corner post and to the soleplate far enough out that you create a triangle with three equal sides.

Another tip: let-in and cut-in bracing adds greatly to the strength of corners, although this type of bracing has lost much of its earlier popularity as newer methods have emerged.

When you are installing top plates and top caps, be sure to lap the timbers so that you do not have 2 × 4s ending at the same place. And be sure to use nails that are long and thick enough to provide real holding power. Do not use any nails smaller than 16d for cap and soleplate nailing.

Use sturdy headers for windows and door, both inside and outside. In many cheaply constructed houses, there are no headers at all, only a 2 × 4 nailed flat over the door opening.

This is a mistake. Headers are neither expensive, difficult, nor time consuming to construct. You can make one in a few minutes and with only three basic pieces of lumber and a dozen or so nails.

Another bad habit is that of using 2 × 4s for all headers. If you use 2 × 4s for headers, use them only in nonbearing walls, because it costs so little to use 2-×-6 headers over all of your windows and doors, you can strengthen your wall without adding to the cost of construction significantly.

You can also add a small amount to the cost by using boards for subflooring rather than plywood, corkboard, or any of the other substitutes for flooring. Plywood is strong enough, but its major advantage might well be the speed with which it is installed. You can nail in a 4-×-8 sheet in moments, whereas subflooring boards require more sawing, fitting, and nailing time. It is worth the extra time and effort, however, to have the strength of diagonal boards tying floor joists together. By extension you add strength to the walls as well.

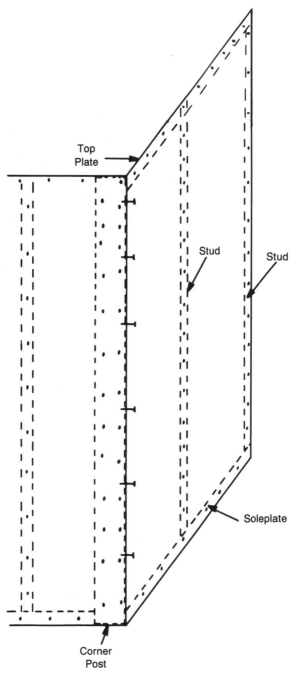

Top
Plate

Stud

Stud

Soleplate

Corner
Post

Fig. 3-4. Plywood panels nailed at corners provide excellent insulation as well as bracing strength against racking.

Take the time to make proper partition junction studs, too. These strengthening additions to a wall cost little, require little time to construct, and add a great deal to the wall. Later, when you tie the nonbearing walls to the bearing walls, you add considerable stability to both walls and insure a good, tight, strong junction of the two walls. Subsequently you will tie all the walls in the house together in the same way, so the effect is repaid doubly in wall strength.

During your framing of bearing walls, do not attempt to save money by scrimping on nails. Use all that are needed to increase the strength of the wall. Do not try to save on studs, either, by using defective ones or by spacing too far apart. 16 inches from center to center is the most effective spacing. You can save a dozen or two studs by cheating, however, you will have weakened the house in order to save $20. It's a poor bargain.

Before you do any framing, take a few minutes—or hours, if necessary—to check, add to, repair, or replace any elements of the foundation, joists, and subflooring. A wall built upon a weak foundation cannot be expected to be as pleasing as one that is properly situated. Chapter Four offers several suggestions for determining that you are ready to start framing.

4

Getting Ready

IF YOU CAN POSSIBLY DO SO, CHECK ALL ASPECTS OF THE BASIC
foundation, joists, and subflooring of your house before you start to frame a wall. If you
have just constructed these elements of the house and know them to be in totally satisfactory
shape, you can proceed to Chapter 5. If you are not completely satisfied that everything
in the foundation and joist and sill area is acceptable, make the following check and make
whatever repairs are needed while you can still do so with a minimum of time and effort.

CONSTRUCTING PIERS

Start by making a visual inspection of the columns or piers under the area where you
plan to frame a wall. Check first to see how far apart the piers are and how they are built.

If possible, you should have some type of support for each floor joist that is at least
twelve feet long, and the support should come as close to the middle of the expanse as
possible. This does not mean that you must build a pier under every joist, however. You
can build a pier under every other joist and use a length of 4 × 4 on top of two piers
so the 4 × 4 supports three joists. To serve the same purpose, you can nail 2-×-4s together side by side and stand them on edge. (See Fig. 4-1.)

If piers are needed, you can build them with bricks or masonry blocks. Before you
start, shovel or dig out the loose dirt and topsoil until you dig into the clay soil. Then
pour in three inches of coarse gravel and build the pier on top of the gravel. You can
buy or pour slabs of concrete for the piers to sit on, if you wish. (See Fig. 4-2.)

If you build the piers on top of loam or loose dirt, however, as the house exerts pressure
on the piers, they will start to sink. Soon they serve little if any purpose, and your joists
will start to sag and bow.

For a good, strong pier, use the following pattern. Mix a batch of mortar and cover
the gravel with two inches of mortar. Then lay one brick so that it is parallel to the joist

and the left side of the brick is one inch out from the right side of the joist. Push the brick into the mortar so that it is embedded firmly. Then lay two more bricks at right angles to the joist so that the ends butt into the first brick. You now have the start of a pier pattern that will be 8 inches by 11½ inches.

Be sure all three bricks are pushed firmly into the mortar and that the cracks between the bricks are all filled. Now change the pattern: lay one brick across the ends of the two bricks that butted into the third, and then butt two bricks into the one that you just laid. Always apply an inch of mortar on top of each course before you lay the next course. (See Fig. 4-3.)

Continue this pattern until you reach the bottom of the floor joist. If you need to, you might wish to use masonry blocks that are thinner than bricks. Or you can either make the mortar joints a little thicker or thinner so that your final course of bricks will be even with the bottom of the joist.

If you wish, use wide boards at the top of the pier to get a proper fit. (See Fig. 4-4.)

For a satisfactory, simple, and workable mortar, use a basic formula of one shovel of cement to three shovels of sand. Start by putting the cement into an adequate container and wetting it. If you have a mixer, put cement and water into the mixer and add the

Fig. 4-1. If you do not want to construct a pier under every joist, you can use one pier and a length of 4 × 4 to provide support for two joists.

sand a little at a time. If you are mixing by hand, work the cement and water together until they are thoroughly mixed. Then add the sand five shovelfuls at a time.

If you use ten shovels of cement, you will need 30 shovels of sand, but don't try to add all of it at once. Add water as needed until all the sand is in the mixture and it is

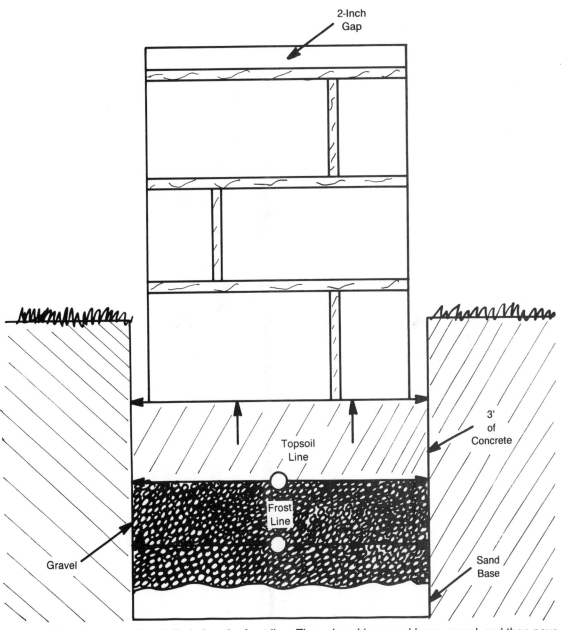

Fig. 4-2. Shovel out topsoil and dig below the frost line. Then shovel in a sand base, gravel, and then pour concrete. Now construct the masonry pier.

thick enough to hold a shape. Use your trowel to cut a trough through the mortar. If it is just wet enough for the trough to remain stable and not start to collapse or seep together, the mortar is usable.

The size of your pier should actually be determined by the weight that is to rest upon it eventually. The patterns given here are for normal weights of bearing walls; if, however, the weight is to be much greater than normal, you can use piers that are in general twice the dimensions of those described.

One problem will likely be that of having the piers end at exactly the right height. As mentioned earlier, you can use bricks of regulation size and then change to thinner

Fig. 4-3. You can make piers of bricks by starting with a base of sand and gravel, then adding concrete. As you build, alter the design of the bricks, and mortar all joints.

concrete blocks—or thicker ones if you need to—or you can simply dig out a little more dirt under the piers and start over. Starting over may be difficult, if you have already mortared the bricks in place, so it is generally wise to lay the bricks in a tentative or trial manner until you are certain the height will be correct. (See Fig. 4-5.)

Remember that when you are laying a test pier that you should allow for about an inch of mortar for each course or row of bricks in any pier. When you reach the top of a pier and see that it is going to be too high for the joist above it, you can always apply pressure on the top two courses of bricks to reduce the thickness of the mortar joint. Or, if the top bricks are going to be lower than you wish, you can thicken the mortar joints slightly. It is better to add slightly more mortar to two or three courses than to make up the needed deficit all in one course.

Once the piers are completed, allow them at least 24 hours of drying time before you put the weight of a foundation or joist on them. As the mortar sets, it becomes harder and harder. Your piers should easily last a lifetime.

Fig. 4-4. Once a pier is built, if it is not the proper height, you can add wooden caps and then saw them neatly for the best appearance.

Don't attempt to save money by using too little cement in your mortar, and don't scrimp on the number of piers you used in your foundation support. It takes only a short time to build a pier, and each one you add can save you money and headaches in the future.

TYPES OF FOUNDATION MATERIALS

The ability of the earth to support a load is called the *soil bearing capacity*. The bearing capacity of soils will vary to a great extent. Topsoil, for instance, will support relatively

Fig. 4-5. Before mortaring bricks, you can stack them to get the basic height. Remember that four mortar joints will equal one brick.

little weight, while subsoil will support a great deal more. Keep in mind that the fewer pounds of pressure or weight per square inch that bear on the foundation, the better the foundation will be. One rule is that a wider foundation will bear weight better than a narrow one, so if you are undecided, build the wider foundation, unless there are prohibitive factors.

Remember, too, that vertical structural elements of a building are considered high-strength members for the simple reason that a piece of wood or other building material that is weak if used on a horizontal manner might be extremely strong if used vertically. One example of this is a three-foot length of 2×4. If stood on end and used as a small column, it will support a great deal more weight than it would if used horizontally between two piers. In the same manner, the 2×4 laid with the broad edge flat will support much less weight than if it is stood on its narrow edge.

The two major classifications of foundations are column, and wall structures, with piers being considered as short columns. In a foundation you may use, in addition to the brick and mortar type suggested earlier, concrete, cut stone, tile, rock, or wood. The best type depends upon the type of building. (See Figs. 4-6 to 4-8.)

Wall foundations are built solid from concrete, cinder blocks, or bricks. The usual rule for width of foundation walls is that each footing should be twice as thick as the wall, and the foundation wall should be the same thickness as that of the wall.

In most modern residential structures the footings, composed of concrete poured over a bed of gravel several inches thick, support a foundation wall of cement blocks or cinder blocks, with bricks often used on the exterior for an attractive appearance.

In some instances the piers in older buildings were made of *coursed rubble*. Such piers, and occasionally entire foundation walls, are composed of rocks, which should be as flat and well-shaped as possible. They are laid without any type of mortar. Rubble foundations can be immensely attractive, from a rustic point of view. To construct a rubble wall, secure a large number of rocks as close to the same size and shape as possible. Dig a footing trench and start laying large flat rocks in as uniform a manner as possible. Each succeeding course should lap the rocks on the earlier course to provide bonding. Such walls are very efficient and extremely durable. Many rural homes still rest upon rubble foundation walls. (See Figs. 4-9 and 10.)

Concrete blocks are often used for three major reasons: they are reasonably inexpensive; they are fast and easy to lay; and they are durable and strong.

An inexperienced mason can lay a concrete or cinder block foundation with little difficulty. After the footings are completed, you apply a generous thickness of mortar to the footings' surface from one end to the other. Then position the first corner block so that it is aligned with both wall lines. Use a masonry line and a square if necessary to ensure proper alignment. Then position the next corner block at the other end of the wall line.

At this point you should stretch a guideline from one block to the other so that you can stay on line without difficulty. To use the guideline, use strong cord tied to a stake, large nail, or spike, and install the nail or stake behind the corner blocks so that the cord must round the corner of the blocks. Check here to see that the cord is in contact with the entire side of the block. If it touches the corner and no other part of the block, shift the block so that constant contact is made. Do this with both corner blocks.

Next use a trowel to mortar one end of the next block to be laid. You can stand the block on end or hold it with one hand and mortar with the other. When the end is mortared, gently lay the block in position so that it makes good contact with the first block and is as close to the guideline as you can position it without actually touching the line. If you allow the block to touch the line, soon the line may be shoved so that the alignment is wrong.

In the same way complete the first row or course of blocks. At each end lay the first blocks of the next wall so you will have a lap for the succeeding courses. Now simply add the blocks needed to complete the wall. Be sure that at the end of each course the last block laps or bonds with the blocks from the next wall. (See Figs. 4-11 and 12.)

If your last block in a course is too long—that is, you did not leave enough space because your mortar joints were too wide—you can tighten mortar joints. If necessary, break a block by tapping it at the proper place with your masonry hammer. Tap in a line

Fig. 4-6. You can pour simple footings for a porch roof or other light building and then set the post atop the pier.

Fig. 4-7. One ancient method of pier building is that of using naturally shaped rocks. Such piers are extremely durable. The one shown here has been in use for over a century.

Fig. 4-8. An easy and inexpensive pier for porches and similar construction can be made by using bricks over a bed of gravel and a 4 × 4 or larger timber.

Fig. 4-9. Rubble foundation walls were once very popular.

from top to bottom of the block until the sound suddenly changes into a deeper tone, then turn the block over and repeat the process on the other side. Usually the block will break in a reasonably straight line at that point. (See Fig. 4-13.)

If the last block is too short, you can make up the deficit by going back to thicken mortar joints. If the space is too wide for mortar joint adjustment, you might have to break a block and use a portion of it to fill the gap.

Solid concrete walls are highly effective. If you choose to construct one, you will need to dig footings and then build wooden forms. Brace these forms adequately. When all forms are complete, have the concrete delivered and poured. If you are mixing your own concrete, however, you can build forms for one wall at a time, tearing down the forms and using them again for the next wall. In this way you can save on lumber. Be sure to allow the walls to set for at least 24 hours before putting weight on them.

Another type of wall, similar to that of coursed rubble, is the *random rubble* wall. This wall differs in that no effort is made to secure a neat, even appearance, and no mortar is used. Such walls have rather short durability and strength and should not be used except for temporary walls.

Tile is rather expensive and wood is highly susceptible to decay and insect damage. Therefore, these materials are not used to construct foundation walls except in special circumstances.

Fig. 4-10. Piers can be made of unshaped stones fitted together as neatly as possible and mortared in position.

If you choose to use wooden piers, dig footings and pour the concrete slabs or lay a bed of bricks or stone for the piers to rest upon. Never allow the wood to come into contact with the soil, and always build the footings for the piers to extend several inches above the soil's surface.

Fig. 4-11. To butter a block, scoop up half a trowel full of mortar and spread it on the block by turning the trowel to the position shown.

Wooden piers should consist of short and thick units of wood, never smaller than 4 × 4 inches, and should be nailed to the joists and braced.

FLOOR SILLS, JOISTS, AND SUBFLOORING

Floor sills are crucial to the quality and durability of your house or the room you are adding. These sills, which are usually no smaller than 2 × 6 or 2 × 8 inches, rest on the foundation walls. The sills are nailed together to form a square or rectangle that

Fig. 4-12. When the block is buttered, lift it and set it so that the new mortar butts against the last block.

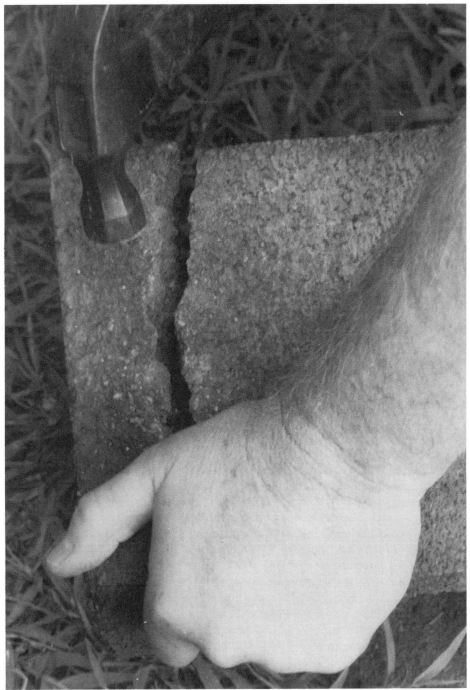

Fig. 4-13. To break a block (to use when there is not enough room for a full block), tap along the desired break line until the tone of the noise deepens and richens. Then turn the block and tap lightly on the other side and the block will break easily.

will represent the exterior limits of the house or room. Years ago very thick sills were used, but nowadays builders generally achieve thickness by doubling the sills.

If you must double sills, don't let two pieces end at or near the same place, if you can avoid it. Butt the ends together. Later if you use a second timber you can determine that the ends are lapped adequately. At the point of the lap use a very generous number of large nails. One extra precaution for both safety and durability is the installation of anchor bolts embedded in the foundation wall with threads sticking up. When you are ready to install sills, bore holes where the bolts appear and then lay the sill so that the bolt protrudes through the hole. Now bolt the sill to the foundation wall for permanent security and protection against high winds.

Sills should be installed with the outside edge flush with the edge of the foundation wall. When they are in place you are ready to nail in the floor joists. Install the outside joist timbers, called *headers* or *joist headers*, on edge. Butt them together at the corners and nail them securely. You now have the sills, installed flat and bolted, with headers installed to make an enclosure to the room or house perimeter.

Now that header joists are in place, mark the places where the joists will be located. Starting at one end of the header joist, place a mark every 16 inches. The traditional method of marking is to locate the sixteen-inch spot, then use a square and pencil to mark the point where the edge of the joist will be located. Usually a large X is made beside the mark so that you will know on which side of the mark the end of the joist will be located. The edge of the joist end will be lined up with the mark and the end of the joist will cover the X.

When one wall is thus marked, go to the opposite wall and do the same thing. When both walls are marked, you are ready to nail in the floor joists. Do so by butting the end of the joist into the header at the marked spot and driving 16d or 20d nails through the header and into the joist.

If you are building a large room or house, and joists are not long enough to reach all the way, you will want to lap them on top of the girders. Lap at least 6 to 10 inches and nail securely.

When all joist are in place, you need to add either cross bridging or solid bridging. Bridging's main purpose is to hold the joists plumb or vertical and in perfect alignment. A secondary purpose of bridging is to distribute heavy weight, such as that of a piano or wood stove, evenly over several joists. (See Fig. 4-14.)

Without adequate bridging it is possible that time and weight will cause slight warping and twisting of the joists, resulting in sagging floors.

Solid bridging consists of short lengths of 2 × 6 or 2 × 8—the same size as the joists—nailed between joists at right angles every 6 feet, or at midpoints in short joists. These bridging pieces can be staggered for easier installation, but they should be kept generally in a line across the width of the expanse.

Cross bridging is usually considered more effective than solid bridging. To construct cross bridging, cut lengths of 1-×-3, 1-×-4, or 2-×-4 lumber and nail them diagonally inside the joists.

To cut accurate length and angle cuts, use a square or chalk line and square a section of joist area 12 to 15 inches long. Use the same length for all future bridging cuts. When you have squared the section, lay a length of bridging stock so that it lines up perfectly with the marks on the tops of the joists. Then mark the stock on the underside and against

Fig. 4-14. Bridging such as the type shown will keep joists from bowing. One strut should be nailed at the top and the other at the bottom edge of the joists.

the inside edge of the joist. Be sure the bridging stock is positioned as it will be nailed between the joists. In other words, don't lay it in a horizontal position to mark it and then nail it in vertically, or vice versa.

Cut along the marks, then nail in the bridging. Do so with 16d nails and position the bridging so that it forms an X.

Some builders prefer that the X be flat, or horizontal, while others like the vertical X. It is not important which style you choose, as long as the bridging is done properly in other respects and at appropriate locations.

When joists and bridging work are complete, you are ready to nail down the subflooring, if you choose to use it.

You have several choices of subflooring. You could use plywood, which would speed up your work greatly, or you could use traditional one-inch subflooring boards. Another choice is to use 1¼-inch treated boards and not use subflooring at all. Or, if your budget will permit, you can use 2-×-6 boards and eliminate subflooring. There are good arguments to support whatever method you plan to use, and time and money figure prominently in all of them.

If you use plywood, the cost may be only slightly higher than the cost of boards. The cost is offset, however, by the fact that you can cover an immense amount of floor space in a matter of minutes, and with only a few nails. If the room's corners are not square, however, you will have trouble keeping the edge of the plywood lined up with floor joists. You may have to trim the entire 8-foot side of the plywood in order to fit it over the joists.

The subflooring boards are tongue-and-groove lumber and the cost is reasonable, the work is neat, and the fit is flawless. You have a great deal of sawing and nailing to do, however. If you elect to use the wider and thicker flooring, the cost per board is high, and you must have a nearly perfect fit or the floor will have cracks in it.

The major argument for using subflooring is that the extra layer of flooring runs diagonally or at right angles to the final flooring, adding greater strength and stability to the floor and, consequently, to the whole room. The double flooring also adds insulation and insect-proofing.

Regular 6-inch subflooring boards are available that can be nailed in place fairly quickly. Ask for No. 2 normal subflooring if you need to buy the lumber. When installing it, you may want to run it diagonally. You can start in one corner and work your way across the room. Or, if you prefer, you can start in the center and work toward the corners, if you are using straight-edged lumber.

Tongue-and-groove lumber is started in the corners of the room. Choose a corner of the room and measure and cut the triangle of lumber that will fit into that corner. Leave the tongue edge to the outside, and when you cut the second and all later pieces, lay the stock against the tongue, so that the groove is aligned. Using only your hands, if possible, force the wood together so that any cracks are closed completely. If you must use added force, lay a section of 2 × 4 gently with a small hammer until it is in place. (See Fig. 4-15.)

As you move into the room so that longer lumber is needed, put two nails in each end of the board and two more where the subflooring crosses joists. If you nail subflooring at right angles to the joists, put two nails again at either end, and at the point where the flooring crosses the joists.

Fig. 4-15. Never hit tongue-and-groove lumber with a hammer. Instead, tap a block of wood against the lumber to keep from damaging it.

One of the major advantages to using straight-edged boards, rather than tongue-and-groove lumber, is that straight boards can be installed so that small cracks are left between them. There is a great chance that your subflooring will be wet repeatedly before the house is dried in, and wet lumber will expand. When the boards expand, they will buckle if there is not an expansion crack left for that purpose. A worse problem is that any studding that has been installed can be pushed out of line by the swelling.

Fig. 4-16. When wet tongue-and-groove flooring swells and buckles, you can saw the tongue off and then hammer the boards back into place. Nail them securely at this time.

If tongue-and-groove subflooring expands and buckles, you may have to run a small-kerfed saw (a circular saw blade is fine) down the crack and saw through where the tongue and groove are joined, and then renail the edges of the boards. (See Fig. 4-16.) Even with straight-edged boards, it is a good idea to leave a ½-inch space between two boards every 6 feet.

When you are using tongue-and-groove flooring, it is a good idea to buy a small *punch*. This small and very inexpensive tool is very useful in nailing flooring down. With proper use you will never need to damage the edges of boards or the tongue or groove of any piece of lumber. If you don't use it, and if you smash a tongue, the groove of the next piece of lumber will not fit well, and you will spend a great deal of time and energy correcting the problem.

To nail tongue-and-groove flooring, position the board where you want it, with the groove on the inside and shoved over the tongue of the preceding board, so that the only nailing surface is the point where the tongue joins the edge of the board. Start your nail at an angle at that point (using cut nails) and drive the nail in until the head is almost flush with the wood.

Now use the punch. Lay it on its side with the large end flush against the nail. Now you can hit the punch edge, forcing the nail into the wood fully and never damaging the wood. If a nail proves to be especially difficult when it is sunk nearly flush, you can use the point of the punch to drive the nail the rest of the way. (See Fig. 4-17.)

Using this technique, your subflooring can be nailed in with no nail heads showing at any point. This will be helpful later when you start to add the final flooring.

At this time you should be ready to start building the exterior walls, assuming you have determined that the footings, foundation walls, headers, joists, bridging, and

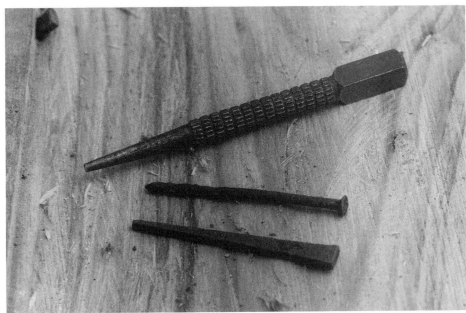

Fig. 4-17. To install tongue-and-groove lumber, use cut nails (such as the one at bottom) and use a punch to sink the head below the wood surface.

subflooring have all been installed to your satisfaction. If you have any doubts, double-check and make the necessary corrections at this point. You can install additional piers later, if you think they are needed, but no vital changes in the foundation can be made after subflooring is installed.

Your next step is that of starting the exterior walls—after you have secured the needed lumber, nails, and tools.

Chapter 5 tells how to install soleplates or bottom plates, determine the studding positions, and frame windows and doors.

Starting the Exterior Wall

BEFORE YOU START CUTTING AND NAILING FOR THE EXTERIOR WALLS,
you should have ready for use the proper building materials. You will need a number
of 8-foot 2 × 4s for studding; the number will depend upon the size of the room and
the space between the studs. The usual spacing is 16 inches, although some area building
codes permit 20-inch spacing, and a few allow even wider spaces between studs.

Because it is your house or room, it is wise to use the best possible building practices.
In this case the 16-inch spacing is best.

CONSTRUCTION MATERIALS

Assume the room is to be 16 × 14 feet. You will need a minimum of about 50 eight-
foot-long 2 × 4s. In addition, you will need longer 2 × 4s for the soleplates and top
plates. Typically, you will need six 16-foot 2 × 4s for the plates and top caps and six
14-foot timbers for the same purpose for the shorter walls.

This might appear to be an excessive number of timbers, but you will be surprised
to learn just how many are needed, if the room is to be built properly. The figures here
include the use of a few 2 × 4s for temporary bracing, some that will be damaged by
careless handling, and some that will be cut incorrectly and will be used for shorter units,
such as cripple studs. (See Fig. 5-1.)

You will also need lumber for header construction. While many builders use 2-×-4
lumber for all interior headers, others prefer to use 2-×-6 or 2-×-8 lumber for this purpose.
Allow for two 44-inch sections for each doorway, or measure the width of the rough door
opening and then allow a couple of inches for squaring. (See Fig. 5-2.)

You will need all 16d nails. Buy a generous supply, as these nails have an almost
universal use. If you have some left over, there will be dozens of uses for them later.

Fig. 5-1. Each corner post uses three studs, while partition junction studs require two studs and other short lengths of studding. Nearly two dozen studs are visible in this photo.

Fig. 5-2. A header can be made up of 2 × 10s and 2 × 4s or 1 × 4s.

LAYING OUT THE PLATES AND CAPS

The *soleplate* is the 2 × 4 that runs the length of each wall between the subflooring and the studs. The *cap* is the timber that runs across the top of the studs in each wall. Caps and soleplates, in short, border the entire wall frame.

When you lay out the soleplate, find the longest, straightest, and best quality 2 × 4 you have. Choose one that is free of imperfections and knots, and make certain it is not warped or weak in any way. This particular timber will in time support the weight of the wall and all that is above it. The soleplate rests on the subflooring, and the subflooring in turn rests on the sills, headers, and joists. A soft or weak spot in the soleplate could later cause the wall to sag in that spot. In turn this would damage the rest of the wall. (See Fig. 5-3.)

Ideally, but not necessarily, the soleplate should be one continuous length of 2 × 4. If you don't have timber long enough, however, you can join two lengths simply by butting one squared end against the squared end of another. If you must join lumber, do so in the middle portions of the soleplate area, not at ends. (See Fig. 5-4.)

If you have 2 × 4s that are long enough to reach all the way across the room, simply lay the timbers in place to be certain that they are straight enough. Lay all four exterior-wall soleplates in place, square the ends, and cut them to their proper lengths. Lay the long wall plates out first, and then locate the plates for the shorter walls. Let the ends of the shorter plates lap over the plates for the long walls.

When you are satisfied with the timbers and their locations, use a square to mark the timbers and then cut them carefully. One note: if you plan to use half-inch sheathing, cut the plates so that the long ones are one-half inch shorter than the actual wall. (See Fig. 5-5.)

In other words, allow a half-inch recess all around the entire perimeter of the subfloor surface. Do not nail any of the plates in place yet. Instead, lay the plate for the first wall flat, and in its proper location. You will need to mark plates for all timbers that will be nailed to it.

Next, use a measuring tape to locate the exact points where studs and doorways will be placed. The studs will be on a 16-inch center, and door openings will vary slightly, depending upon the width of the door. (See Fig. 5-6.)

When you mark the stud locations, make a rectangle with an X inside it so there will not be any question later as to where the stud will fit. When you have completed marking the plate, you can place another 2 × 4 beside each of the soleplate timbers and mark it for studding exactly the way you marked the first one. This will be the top plate. (See Fig. 5-7.)

An easy way to do this is to lay the timbers flat. Use a straightedge to extend the stud markings on the soleplate to the next 2 × 4, which will be the top plate.

It is important for all studding marks to be aligned perfectly. If they are not, studs will be installed in a crooked fashion, and you will experience great difficulties when you try to install wall coverings. (See Fig. 5-8.)

Don't worry about marking door openings for the top plate, because door frames will not extend that high.

If a top plate joint ends on top of a stud, you will want to change it to let it end between studs. (See Fig. 5-9.)

At the end of the soleplate and top plate mark the location of the corner posts, and allow 5 inches for each. Chapter 6: Installing Corner Posts, will show you how to construct these posts.

One more checkpoint: if you have to join 2 × 4s when you are laying out the soleplate and top plate, be sure that the joints do not end at the same place. If this happens, your wall will be weakened considerably. If you have already cut the timbers, you can simply reverse the top plate and, unless they are exactly the same length, the joints will be staggered, at least slightly.

At this point you have laid out the soleplate and top plate for one exterior wall. You have also marked the locations for the studding, the door frame, if any, and the corner posts.

If there is to be a window in the wall, mark the spot where the exact center of the window will be. Then measure and mark one-half the width of the window on each side of the center mark. Using a heavy felt-tipped pen or pencil, mark the window points clearly, since you will want to put cripple studs rather than full length ones in the space. (See Fig. 5-10.)

If you are fortunate enough to have helpers, you can plan to complete the construction of the exterior wall as it lies on the subflooring. If you are working alone, you need to

Fig. 5-3. Use your longest, straightest, and best quality 2 × 4s for soleplates. It is better to have an uninterrupted timber reach completely along a wall frame.

locate some temporary assistance if you plan to lift the entire wall into its position and nail it in place. The completed wall frame, while constructed of lightweight woods, is too heavy for one person to lift. Don't try lifting the wall frame alone, because you would have to hoist it into position, hold it in place while nailing the soleplate to the subflooring, and support it in a perfectly vertical fashion.

Fig. 5-4. Lay out all wall frame timbers, including both soleplate and top plate, partition studs, corner posts, cripples and headers.

COMPLETING THE STUDDING

If you have assistance, you are now ready to nail in the studding and window and door frames. The simplest way to nail in the studs is, first, to cut all studs to the exact length you will need for the wall frame. Then, after spreading the soleplate and top plate 8 feet apart, position the studs so that one end is in its exact location against the soleplate. Use 16d nails and nail through the bottom of the soleplate and into the stud. Be sure that the nails are sunk completely and that the end of the stud is seated evenly against the soleplate. (See Fig. 5-11.)

Now go to the top plate and repeat the process. This time, when the end of the stud is in its precise location, nail through the top of the top plate. Use 16d nails and make certain that the end of the stud is seated evenly and that all nails are sunk fully. (See Fig. 5-12.)

It is easier to complete the nailing if you lay a 2 × 4 under the assembly, so that the wall frame is 2 inches off the subflooring. In this fashion the assembly will be off the floor far enough to permit space for hammering.

After you have nailed in the studding, start to build the headers for doorways and windows, and cut partition studs and cripple studs. As with all construction, it is necessary

Fig. 5-5. If you plan to face the wall with brick, build a wall frame recess sufficient to accommodate diagonal bracing without interfering with the brick line.

to have exact measurements. Take plenty of time when you mark and cut the studding, unless you bought precut lumber. Don't mark and cut one cripple stud, then use the stud as a pattern for the next cut, and then use the second stud as the pattern for the third. You would be lengthening each stud by the width of your pencil mark plus the space needed to get the pencil into position against the end of the pattern piece. It is better to cut one piece, measure it a second time to be sure that it is correct, and then use that first piece

Fig. 5-6. The markings on the soleplate and top plate show the door stud locations as well as junction post locations.

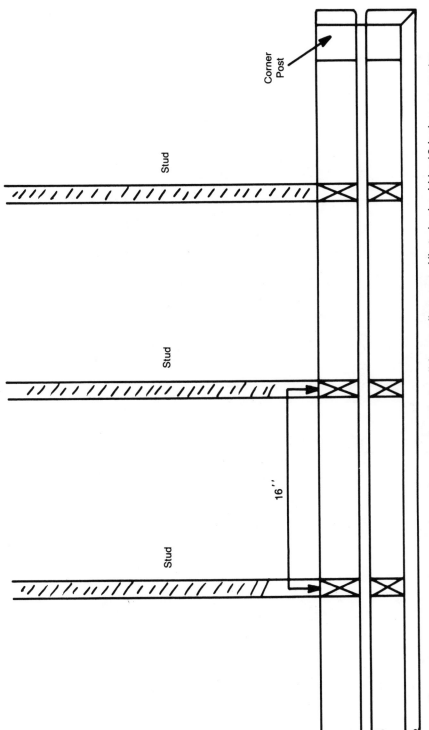

Corner Post

Stud

Stud

Stud

16''

Fig. 5-7. Mark both plates at the same time to be sure of perfect wall frame alignment. All studs should be 16 inches on center.

75

as a pattern for all future cuts. This holds true not only for studding but for all types of cuts you will be making.

When you have nailed in all the studding except for the door and window cripples (a *cripple stud* is a short one that is used over and under window openings and above doorways), you can nail in the corner posts. (Refer to Chapter 6 for directions on constructing corner posts.)

Now use a tape measure to determine that the wall is as square as you can get it. Measure from the outside corner of the top to the outside corner of the bottom, diagonally.

Fig. 5-8. When both plates are marked, place the top plate in position and nail in studs by sinking nails from the outside edges of both plates into the ends of the studs.

Fig. 5-9. Do not let a top plate joint occur over a stud. Locate the joint between studs and, if you choose, use a support length of 2 × 4 nailed under the joint.

Fig. 5-10. Cripple stud locations should be marked the same as common stud locations. Nail these in place before raising the wall.

Then measure from the other top corner to the bottom of the opposite corner and compare results. The distance should be exactly the same. If it is not, make corrections at this point.

You can start by using a square on each corner. If the corners are not square, you will need to determine where you went wrong and make necessary adjustments. It is quite possible that the entire assembly was allowed to shift during the building. In that case, all you need to do is use a pry bar to shove one corner into its correct position.

If that fails, you will have to knock the top plate loose and reposition it and the studs that are nailed to it. If you are off only a fraction of an inch, you are close enough; anything more than an inch should definitely be corrected.

Fig. 5-11. You can buy studs that are pre-cut, such as the ones shown here. You do not have to cut the studs to exact lengths, and the studs are slightly cheaper than traditional 2-×-4 lumber 8 feet long.

INSTALLING HEADERS

Headers can be made in any of several ways, but the major consideration is that the header will be the same exact thickness as the door framing. Some builders simply cut three lengths of 2 × 4 and stand two on edge, side by side, then lay a third one, flat, on top of the first two. Make sure that the edges of the top one are flush with the outer edges of the two bottom ones, and then nail the assembly together. (See Fig. 5-13.)

Fig. 5-12. Note that the ends of 2 × 4s in this wall frame are inside the door frame. This is a perfect place for them, because that portion of the soleplate will be removed later.

If you need extra width, use two or three strips of plywood or one-inch boards, three inches wide, and nail these between the two 2 × 4s standing on edge.

For a more substantial header, use 2-×-6 or 2-×-8 lumber for sides and a 2 × 4 for the bottom. Nail these together in the same way, with the 2 × 4 serving as the bottom. The width of the opening will determine the type of header you need most.

Fig. 5-13. The usual length of a door header is 44 inches, which allows for installation of trimmers and jambs later. You can organize your work well by making all headers at one time.

If you are building headers for a window or normal-sized interior door, the 2-×-4 headers will probably suffice. If you have double windows or French doors, or any type of double doors, however, you will want to use at least 2-×-6 lumber for headers.

If you want to eliminate the time and energy needed for building the headers, you can use 4-×-4 lumber, although you will find this material a little more expensive.

Keep in mind that all headers are approximately four inches longer than the widths of the rough door or window openings. You will need to add the door trimmers.

Allow about seven inches of space between the door headers and the ceiling. Headers for windows are even longer. These headers are usually nailed between the studs on either side of the windows, so the headers will be considerably longer than the width of the rough window openings. A four-foot header is generally the right length for the average-sized window.

The purpose of the headers is to provide the greatest amount of support for the walls above the door or window openings. The headers work in conjunction with the sills at the bottom of the window openings to lend stability to the walls.

You can nail the headers in while the wall assembly lies on the floor, or you can install them after the wall is erected. One way is about as easy as the other. Many people find that it is a very simple matter, once correct measurements have been made, to nail in headers in flat walls, because everything can be reached and held easily. (See Fig. 5-14.)

Here is one easy way to nail in a header. Decide how great the distance between header and top plate will be. Cut a couple pieces of scrap wood the proper length and tack these to the studding with one end tight against the top plate. Now, with the header pushed firmly against the scrap wood, nail the header in place by angle-nailing or toe-nailing with a 16d nail from the header into the studding. The length of scrap wood will prevent the header from moving up too far while you nail it. When the header is installed, you can pull the scrap wood out and discard it or save it for use in another door or window.

You should be sure to double the studding on both sides of all windows and doors in order to provide the maximum support for the wall at these points. When the header is in place, you need to nail in cripple studs. These short studs are spaced, as are regular studs, on a 16-inch center. If the window is of such a narrow width that only one cripple is required, however, it can be nailed in the center of the window. When possible, though, maintain the basic centering policy for all studs. (See Fig. 5-15.)

This is true for cripples under windows as well as for those above both windows and doors. The reason for the 16-inch center for cripple studs is obvious. The spacing for a three-stud series equals 48 inches, and nearly all paneling and wallboard measures 48 inches wide. Wall coverings will always end in the center of a stud, therefore, in a wall with 16-inch stud spacing.

Although your cripple studs may be short and, from all basic appearances, unimportant, do not use flawed timbers for this purpose. Remember that windows, especially double-wide ones, are the largest expanse of minimally supported wall space. Therefore, use sturdy headers that are nailed in place solidly and use sound and strong cripple studs. These members are of genuine importance to the strength of the wall. In a double-window wall space, there is an 8-foot gap of empty wall. You would never dream of spacing studs in a solid wall eight feet apart, but this is precisely what you do if you do not use the best-constructed headers and the best cripple studs available.

Remember that the window sill is the bottom equivalent to the header, and sills are equally important. Don't hastily nail in a short piece of inferior scrap wood simply because you don't want to cut into a perfectly good eight-foot stud to produce a sound window sill.

RAISING THE ENTIRE WALL

When the entire wall frame is completed and ready to go up into position, first of all, secure assistance. Do not risk a serious and possibly permanent injury to your back by attempting to lift the entire assembly.

Fig. 5-14. You can install headers by angle-nailing or toe-nailing the header between studs.

Fig. 5-15. Here is a properly supported window frame. Note the large header, cripple studs, and partition post.

Before the wall frame is ready to be erected, you should have constructed rough openings for windows and doors. All common studs, cripple studs, headers, and corner posts should be nailed in place.

Before you try to lift the assembly, you need to raise it about four inches off the floor so everyone can get a good grip under the top plate. One of the easiest ways to do this is to hit the top plate, about three feet from the end, with a sharp blow from the claws of your hammer. The claws will penetrate the soft wood of the top plate, providing an easy lifting point.

As you, or the designated lifter, pull the assembly slightly off the floor, someone else can slip a short piece of 2 × 4 under the corner. Move to the other end of the wall frame and do the same thing. The entire frame is thereby raised off the floor and is easier to lift.

Before the actual lifting of the frame, go to a point about 7½ feet from the edge of the outside edge of the room. Nail to the subflooring a 2-foot section of 2 × 4 that is parallel to the studding. Do this just inside the corner posts on each end of the wall frame assembly.

Locate two long, sturdy timbers (2-×-4 or 1-×-4 boards) and lay them close at hand. Also have a hammer and nails ready to use quickly. These timbers are braces to hold the wall frame in place once it is lifted.

If you have three or four people working with you, all four people can lift the wall frame to an erect position. Raise the wall to as perfectly vertical a position as you can manage without a level. One person then should leave the wall frame assembly and nail one of the brace timbers to the inside top of the corner posts on each end of the framing. (See Fig. 5-16.) As soon as one brace is nailed to the corner post (at the top and on the inside), the nailer should nail the other end to the short 2 × 4 (parallel to the studs) that you nailed to the subflooring earlier. (See Fig. 5-17.)

Now one side of the wall frame is fairly secure. Next, repeat the bracing on the other end. At this point the wall will stand by itself and the lifters can turn their attention to other important tasks.

The first task is that of positioning the bottom of the wall frame in its exact and permanent location. It should be perfectly parallel with the outside edge of the wall and one-half inch from the edge. One easy way to determine the exact location is to mark a chalk line before the wall frame is lifted.

The frame might need to be shifted slightly. To do so, tap the soleplate with a large hammer until the frame is in its precise spot. (See Fig. 5-18.) You can then drive 20d nails through the soleplate, two or three between each set of studs, and into the subflooring. Each nail that goes in stabilizes the wall frame a little more, so be generous in your use of nails.

When the soleplate is nailed solidly to the subflooring, you can make whatever adjustments need to be made to get the wall frame assembly to a vertical position. To do so, first set a level on the corner posts to see which way and how far the top needs to be moved. You know that the bottom is in place, so any adjusting will have to be done at the top. If the corner posts are not vertical, tap the brace timber loose at its subfloor mooring. Let one person hold it ready to be renailed as soon as the wall is straight. Another person holds the level on the corner post, while another person moves the top of the wall frame slightly and slowly until the exact position is achieved.

Now renail the brace timber to the anchor point, and move to the other end. Do the same thing there, and at this point the first wall frame is permanently positioned.

The wall frame will have to stand, unsupported except for the nails in the soleplate and the brace timbers, until another exterior wall section is completed and nailed into place. Therefore, if the weather is stormy, or if high winds are prevalent in your area, you may want to add more brace timbers. A severe gust of wind can blow a wall frame into the

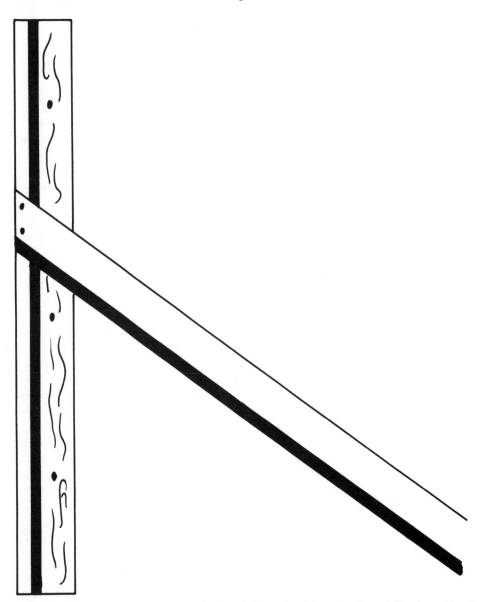

Fig. 5-16. Before the wall frame is raised, nail a length of 2 × 4 to the subflooring several feet away. When the frame is raised, run a brace from the frame to the timber nailed to the floor.

yard. Not only is there physical damage, and time lost in repairing the work, but the damaged frame also presents a safety hazard.

If the wall frame is assembled and ready to erect only a short time before the end of a day's work, it might be wise to leave it lying down and raise it first thing the next day. In this way, the wall will not have to stand relatively unsupported through the night.

As soon as possible start construction on the next wall, obviously one that will tie into the one just erected. Again, lay out the soleplate and top plate, mark these for corner posts, windows, doors, and all types of studs, and cut studding and headers. Prepare all the timbers so the next wall unit is ready to erect.

Assemble the wall, raise it as you did before, and nail in the timber braces. Repeat the entire earlier process, and when the wall is vertical and braced and soleplate nailed to the floor, you will see the importance of the corner posts, because now you can nail the two corners together and your wall has dramatically increased in strength and sturdiness. (See Fig. 5-19.)

Following the same work pattern, assemble and erect, then nail together at corner posts, all outside walls. When this is done, you are ready to assemble interior walls.

FRAMING A WALL UNASSISTED

Very few jobs are easier to do alone than with assistance. Wall framing is not among them, but it can be done. Working alone is not as bad a handicap as it may seem at first glance.

Fig. 5-17. If you do not have subflooring installed, you can drive posts into the ground and attach the braces to the posts. If you prefer, you can run braces to adjacent sills.

Fig. 5-18. If the bottom edge of the soleplate is out of line, hit the top plate with a hammer to drive the frame into its proper position.

You can easily construct headers, nail in studs, and do a great number of the basic tasks of wall framing alone. There are a few jobs, however, that may require some insight as to how to accomplish them.

You can accomplish your first task, that of chalking a line for the soleplate, without trouble. Simply tap a nail lightly into the subflooring at the extreme end of where the

Fig. 5-19. If other wall frames are erected, nail the newly raised wall to the existing corner posts or studding.

wall will go, and one-half inch from the outside edge. Then loop the chalk line around the nail. Secure this line while you pull it to the other end of the wall line, locate the half-inch mark, and snap the chalk line.

The remaining wall construction tasks can be handled in one of two ways. The first method is as follows. Place and mark the soleplate and top plate and nail in corner posts and alternate studs. Depending upon your strength and the length of the wall, you can nail in all the studs and make the rough door and window openings. In short, put together as much of the wall as you can lift and hold without difficulty. It may be that you will want to nail together only the sole and top plates and corner posts.

Whatever part of the wall you elect to try, your major job is to lift the assembly and hold it steady while you get brace timbers nailed securely. As difficult as it may sound, you can do this working alone. One of the best methods is the following: nail the *scotch block*—the short 2-×-4 length where the timber brace will rest on the floor—in place, preferably at the middle of the wall. You might need to make a quick estimate of the length needed. The best way to do this is to rest one end of the brace timber against the block on the floor. Hold the timber at the approximate 8-foot height so that you will know for certain that the timber will be long enough for your purposes.

Slip the end of the timber under the soleplate and allow the other end to rest close to the top of the middle stud. Using one 16d nail, nail the top end of the timber to the middle stud. Do not nail it so tightly that the brace can't move freely. Lift the wall assembly, and the end of the brace timber that was under the soleplate will swing free. Move this free end out to the block you nailed to the floor earlier. Instead of nailing the timber to the block, let it rest against it. The weight of the wall pushing the brace against the block will hold it in place.

Your wall might not be perfectly vertical, but you can nail in scotch blocks at either end of the wall and nail more timber braces to them. After you have nailed one end of the brace to the floor block, start a nail in the other end. With your hammer in your belt or hanging from a belt loop, and with the level in your other hand, go to the corner post where you are working and shove the corner post gently with your shoulder while you hold the level against the post. When you find a proper reading on the level, hold the post steady. Position the timber brace, which you have held against the corner post with your body pressure, against the corner post. Using your free hand, drive the nail you started into the post. Now one end of the wall is vertical.

Do the same at the other end. Nail the brace in place, and you are now ready to nail the soleplate to the floor. You can nail the remainder of the studs, if any, in place with the wall standing erect. You will have to toe-nail them instead of nailing through the bottom of the soleplate, although you can still nail the studs through the top of the top plate.

The problem with toe-nailing is that the stud has a tendency to move as you hammer the nails. You can overcome this small problem, however, by measuring the distance from the right side of the first stud to the left side of the corner post and then cutting a small length of 2 × 4 that will fit exactly between stud and corner post. When you start to nail, lay the length of wood on the soleplate and it will keep the stud from slipping. You can do the same at the top by using a small C-clamp to hold the length of wood in place.

If, because of doubled studs, the distance is not the same and the length of wood will not fit, you can cut another length of scrap wood. For another solution, fasten the C-clamp tightly over the soleplate and against the stud, and the clamp will hold the stud in place while you nail.

Here's a small trick for nailing in the heavier headers. First, select a 2-×-4 piece about eight feet long. Next, measure the distance from the subflooring to the point where the bottom of the header will be when it is in its proper location. Mark that point on the 2 × 4 and nail a short piece of scrap wood to the 2 × 4 so that the top of the scrap wood aligns with the mark corresponding to the bottom of the header.

Make two of these and clamp or nail them to the inside of the studs where the header will be nailed. Lift the header into place, letting it rest on the scrap wood nailed to the 2 × 4s, and you can nail it easily in place. Similar devices will work for other nailing problems.

The second method of wall framing unassisted is perhaps even easier. Start by chalking the line for the soleplate and then nailing the plate to the subflooring before you add one timber to the assembly. With the soleplate in place, locate the corner posts and nail them in place, by the following method. Mark the spot where the corner post will be located on the soleplate. Nail a piece of scrap wood so that the edge of the soleplate will be against the scrap wood when the post is placed on the soleplate. Next, nail one end of a timber brace—a short one this time, no more than 5 feet long—to the corner post about halfway up, and so that the brace can extend alongside the soleplate.

When you lift the corner post into place, position it so that the edge is against the block you nail to the soleplate. Then extend the timber brace until it will support the corner post in a nearly vertical position. Drive two 20d nails at an angle through the corner post and into the soleplate, then use your level to get the post perfectly vertical. When you have done so, you can tack the timber brace to a block on the subflooring. This brace will hold the corner post in perfect position while you complete nailing it in place.

Do the same with the other corner post. Then gently lay the top plate into position atop the posts and nail it into place. You have already marked the stud locations, and you can proceed to nail in studs, headers, and the rest of the wall. You can nail in rough door and window frames easily at this point.

Before you nail in any studs you might want to use a square and check to see that all four corners of the wall framing are correct. If you wish to be doubly certain, you can nail in short braces from posts to soleplate to make certain that the perfectly squared corners remain that way. You then can nail up timber braces to hold the completed wall frame assembly steady until other wall sections can be framed and put in place.

Throughout this chapter numerous comments have been made about the all-important corner posts. Chapter 6: Installing Corner Posts will explain how to construct various types of corner posts and will offer more suggestions concerning how to install the posts effectively.

6

All About Corner Posts

THE CORNER POST IN A WALL FRAME HAS SEVERAL FUNCTIONS.
Each post supports a major part of the wall and, as a result, all that is above the wall, including attic joists, rafters, roof, and possibly other rooms. The corner post stabilizes the corners of the room, the point where some of the greatest stress is placed. It also helps to keep the entire wall vertical and true and the entire room square. Finally, it provides the necessary points where the other walls can be joined at maximum strength.

TYPES OF CORNER POSTS

Because of the importance of corner posts, you should not attempt to save on materials or money by making short cuts. All corner posts should be constructed as sturdily as reasonably possible. Nothing smaller than 2 × 4s should be used.

To construct one type of corner post, use a 4 × 6 with one 2 × 4 nailed on the side where the siding will be added later. The 2 × 4 should be flush with the corner edge of the timber. Such a corner post will provide an immense amount of strength and sturdiness. However, because of the cost of 4-×-6 timbers, many builders prefer or are forced to use smaller lumber. (See Fig. 6-1.)

A good second choice is a 4 × 4 with 2-×-4 studs nailed to the two sides that face into the room you are framing.

For all outside corners, it is wise to use nothing smaller or weaker than the two types of corner posts described above. If you need to use smaller timbers, you can use 2 × 4s to make fairly sturdy corner posts for interior use. (See Fig. 6-2.)

One widely used type of corner post is made of two studs that are nailed together, with the 2-inch side of one nailed against the 4-inch side of the other so that a right angle is formed with an uneven V on the inside of the post.

A fourth type of corner post is composed of three studs and three foot-long blocks of 2 × 4. First, lay one 2 × 4 flat on the floor and nail the foot-long block to it, flush. One block should be nailed along the top of the stud parallel with it, and with the ends even. A second block is nailed similarly at the bottom, and the third is nailed in the center. A second stud is then nailed over the blocks parallel with the first stud, and a third stud is nailed flush with one edge of one of the original studs.

If you are working alone, complete the corner post and nail a set of braces to it so that you can nail it temporarily in place. If you are working with assistants, you can nail the corner post in as you complete the wall-frame assembly.

To build all corner posts, use the best quality studs or other timbers you have at your disposal. Reject all timbers that are weak, warped, or defective in any significant way.

When using 4-×-6 corner posts, nail in the posts with 20d nails—never anything smaller. Use four nails if you are nailing the post in as part of a complete assembly; that is, if the wall frame is lying flat and you are nailing into the corner post through the bottom of the soleplate and top of the top plate. If you nail it in separately, you will have to toe-nail or angle-nail it in place. In this case you should use two nails on each side of the post at the bottom and be sure to drive them in firmly. Repeat the process at the top. Such a timber is rather heavy and needs to be held securely in position.

When you are using smaller timbers, such as 2 × 4s, you can use 16d nails to nail the studs together. Space nails no more than 1½ feet apart, and nail down both sides of the 4-inch side of the studs.

Fig. 6-1. The simplest corner posts are those composed of a 2 × 4 nailed flush to a 4 × 6. Most builders prefer to use the smaller studding, as shown.

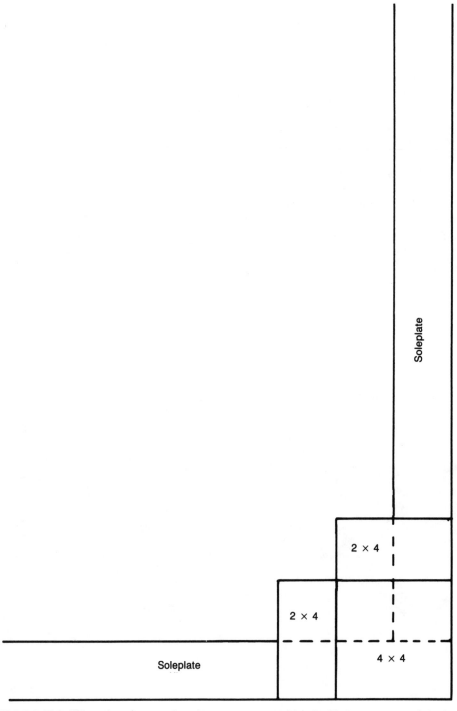

Fig. 6-2. This illustration shows a 4-×-4 corner post combined with 2 × 4s on each inside surface.

CONSTRUCTING T-POSTS

A T-post is used wherever a partition wall of a room meets an outside wall. The purpose of the T-post is to provide an adequate nailing surface to use when you are installing wall coverings. You can make a T-post in any of several ways.

One of the easiest ways is to lay a 2-×-6 stud flat on the subflooring and center a 2-×-4 stud upon the first timber. Position the 2 × 4 so that you have exactly the same space on both sides. Nail the 2-×-4 stud in place by using 16d nails spaced no more than 18 inches apart and in two rows.

One rather expensive way of constructing a T-post is to nail two 4 × 4s together. Then center 2-×-4 studs the way you did in the previous paragraph. The difficulty here is in nailing the 4 × 4s securely together, since they are rather thick and long nails are required. You should space the nails 16 inches apart and nail from both sides. Lay one timber on top of the other and nail through the top one, making two rows. Then turn the assembly over and repeat the process so that you have nails actually meeting or nearly meeting each other.

One simple, inexpensive, and very effective way to make a T-post is to lay a 2-×-4 stud flat and nail three 2-×-4 blocks to it, nailing the blocks, as before, flush at the top, the middle, and the bottom. Next, nail a second 2 × 4 in place. Lay it over the blocks and flush with all edges. Use 16d nails spaced no more than 15 or 16 inches apart and in two rows. At this point the blocks are sandwiched between studs, and you can lay a third stud over the assembly. Center this stud so that it covers the blocks. Nail the stud in place as you did previously.

A final method of T-post construction requires one stud, a 2-×-6 timber, and some bridging lumber. Lay the 2 × 6 down and nail a 2 × 4, centered, on top of it. When you are ready to install it you can nail in some bridging between the studs on the two sides of the T-post position. Then nail the T-post to the bridging and to the sole and top plates. (See Figs. 6-3 and 4.)

You may wish to construct partition posts to use where two interior walls meet. To build such posts, all you need to do is modify the corner post construction slightly. You can center a 2-×-4 stud flat on the face side of a 2-×-6 timber and nail it in place. Then turn the assembly over, center another stud, and nail it in place also.

A second method of constructing a partition post is to make the block "sandwich" by nailing three blocks between 2-×-4 studs and then centering a stud over the blocks and nailing it in. Then turn it over and nail in a second stud as you did on the first side.

THE PERFECTLY VERTICAL WALL

It is obvious that perfectly vertical walls are necessary for stable construction. If you are working alone and erecting wall framing by individual timbers, however, it is far easier than you might imagine to wind up with wall frames that are not true. That is, the wall frame assemblies are not perfectly square or vertical.

The same problem might occur if a team of workers erects a wall frame. It is more likely to be true of the lone worker, however, because in a team several pairs of eyes are likely to detect a problem.

Fig. 6-3. To construct an adequate T-post, nail in blocking between two studs. Blocking should be installed at top, bottom, and middle of the stud length.

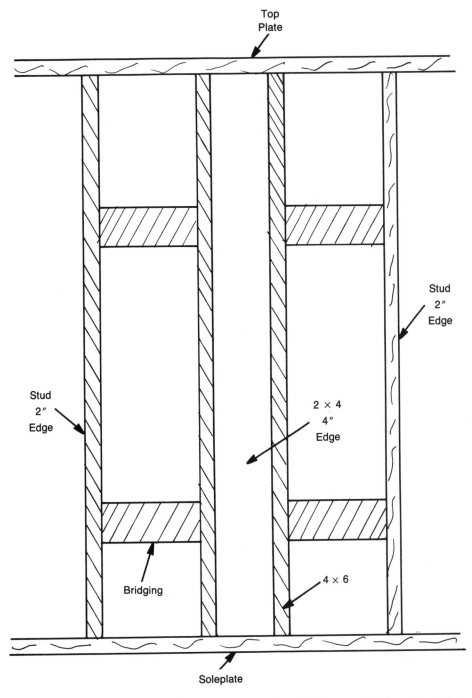

Fig. 6-4. You can also strengthen studding by adding blocking one-third and two-thirds of the way down the studs. Install the blocking by sinking nails from the opposite side of the studs into the end of the blocking timbers.

First, here's what happens if a wall frame is not true. Adjacent or connecting walls will not fit properly. There might be gaps between the walls, or the walls that do not have adequate room.

You might erect a wall frame that leans in slightly, even with only an inch or two variation from bottom to top. If the wall that butts into it or joins it, then, is exactly true, the two walls may fit perfectly at the bottom where the soleplates join. The second wall, however, will overlap the edge of the corner post of the first wall by the inch or 2 inches. Before you can join the two walls, you will have to reposition the first wall or pry it out so that the corner post of the second wall can fit flush against it.

This is much easier to visualize than it is to accomplish, particularly if you are short-handed in assistance. If the corner post of one wall is bowed or slightly warped, you will have a gap where the two walls meet. You might then anticipate severe problems with wall coverings.

In a soundly constructed wall frame, all members of the wall should fit together as snugly as pieces of a jigsaw puzzle. There should be no unnecessary stress on any of the timbers. If you have to use a crow bar to pry corner posts apart in order to join wall frame units, or if you have to pull corners together and hold them until they are nailed, you have a problem that should be corrected before you go any further. Three tools, the level, the square, and the measuring tape, can help you locate and correct the problem quickly.

When the corner post was nailed in temporarily, you nailed it only enough to keep it erect while you secured the temporary bracing. After you adjust the bracing, you can then complete the permanent nailing of the corner post.

One problem might be that the corner post was never completely vertical, even though you checked it with a level and saw that the reading was correct. It is possible that the level was resting on a slight irregularity on the surface of the post and an incorrect reading resulted.

Two or three simple steps will usually prevent such an occurrence. First, when taking a reading from a level, hold the level so that the bubble is at or near eye level. If you are looking up or down at the bubble, your angle of vision could be misleading. One end of the level could be resting upon the head of a nail that was not driven in fully, and a flawed reading might register.

Try the level at two or three points on the corner posts, therefore, and then do the same thing on other sides of the post. It is very possible that the post is perfectly vertical if the reading is taken on one side, but leaning if the reading is taken on another side.

Next, use your square to see if you have true angles where the corner post is nailed to the soleplate and to the top plate. When you try the square on the soleplate, if the instrument does not fit satisfactorily, locate and correct the cause of the problem. Do the same at the juncture of top plate and corner post.

Measure the wall frame diagonally from corner to corner while it is lying on the floor. Measure both ways to be sure you get the same reading. You can also measure from the outside edge of the corner post at the top plate to the outside edge at the other end of the wall frame, and then compare the reading with one taken from the outside edge of the posts at the bottom of the frame. (See Fig. 6-5.)

Even if the corner posts are perfectly vertical after you finish nailing them in, they could be moved slightly by bumping or other subconscious actions. Good bracing, therefore, should be nailed in before you finish the wall frame.

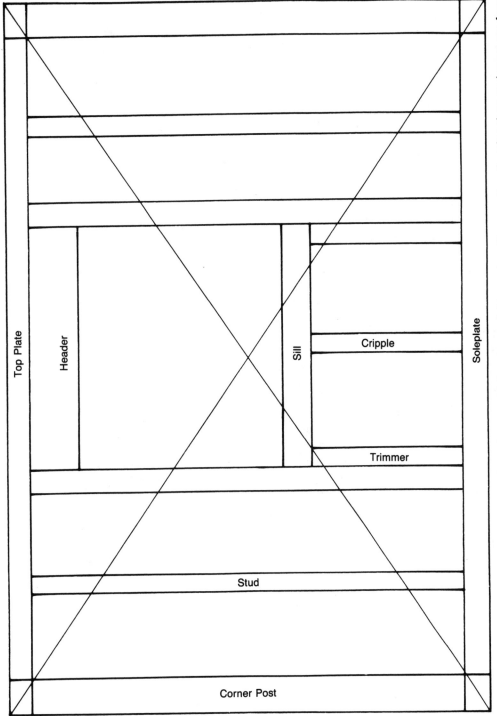

Top Plate

Header

Sill

Cripple

Soleplate

Trimmer

Stud

Corner Post

Fig. 6-5. Measure the laid-out wall diagonally from the upper left to lower right corners. Reverse the direction then and measure from upper right to lower left corners. The distances should be the same, within a fraction of an inch.

Earlier we provided information on cut-in bracing and let-in bracing. Let-in bracing is set into the narrow edges of studs, and it can be done at this point, in one of two ways. If the wall-frame assembly is raised as one unit, the let-in bracing must be done in the following manner. Mark the studs and corner post, cut out the necessary sections, and then insert the brace and nail it in place.

You can mark the studs and corner post in two ways. The first way is to hold a 1-×-4 board in position and mark studs and corner post on both sides of the brace timber. To mark the brace location properly, you may need to hold the brace in place while you nail it temporarily. Do so by sinking two small nails partway (so they can be pulled out easily later), one in the corner post and the other in the final stud that the brace will reach. Now the brace will not move while you mark it.

While you are marking, be sure to mark the outer edge of the corner post and the final stud. Then take the brace timber down and cut the ends where you marked them but nowhere else.

Using a handsaw, cut along the lines on the studs and corner plate, but cut in only enough for the brace timber to fit. You do not want to weaken the studs by cutting too deep. When the cuts are made, use a wood chisel—or the point of a large slothead screwdriver, if no chisel is available—and chip out the wood between the two marks. Now put the brace timber in place and nail it.

A second way to mark the cut lines is to measure the width of the brace timber and then use a chalkline to indicate the marks. You can drive a small nail partially into the corner post, loop the end of the chalkline over it, pull the line to the desired stud, and snap the line. Then measure the exact width of the brace timber, move the nail to the proper point, and make another chalkline perfectly parallel with the first, and proceed as you did before.

If you choose to use cut-in bracing, you can mark the brace lumber easily in much the same way. While let-in bracing means cutting the corner posts and studs slightly, cut-in bracing means cutting the brace timbers diagonally into sections so they will fit tightly between studs. You can hold the brace timber, in this case a 2 × 4, narrow edge against the studs, while your helper marks it. If you do not have a helper, you can mark it alone.

Since the bracing will consist of short lengths of 2 × 4s, this is an ideal opportunity to use some of the scrap ends of studs and other timbers that have accumulated during your work. A piece that is about 20 inches long will suffice, since you will be bracing only between corner post and stud and between studs, none of which should be more than 16 inches apart.

The first step is to mark the line of the brace by using the chalkline with the end looped over a nailhead. Stretch the line and snap it to mark where the braces will go. Then start at the corner formed by the bottom of the corner post and the soleplate. Hold a 2 × 4, with the 2-inch edge against the stud and the other end extending slightly past the corner, so that the bottom of the 2 × 4 is barely touching the chalk line. Hold it firmly while you reach behind the brace and make a pencil mark along the inside edge of the corner post and another against the edge of the first stud.

Saw both ends along the line. When you are finished, insert the brace so that it fits well into the corner and against the stud. Nail it in place.

Repeat the process with the other studs. When you are finished you will have a corner post and wall frame that is well supported, at least on that one side.

Fig. 6-6. Note the diagonal bracing used at various points in this series of wall frames. Note also that the corner posts or junction posts are installed at all future doorways.

Fig. 6-7. When you install diagonal bracing near a wide window or door opening, the bracing should continue in a line from the studding to a cripple stud. In this photograph the bracing has not yet been installed, but it should end at the center cripple stud from each side. It will be interrupted in the actual window space.

When you finish the corner, go to the other corner and brace it in the same way. When you erect the next segment of wall frame, install the corner bracing in the same way. Do this with all corners as you complete the framing.

To add even more strength, you can install 2-×-4 blocking between studs so that the wall is supported diagonally from two directions and horizontally. The stud blocking is put in just as you did the diagonal cut-in bracing, except that you cut the pieces straight across the ends rather than diagonally, and you nail them in as before. When you make the cuts, try to get the closest fit you possibly can. Do not fit the blocks so tightly that the studs are bowed, but do not cut the blocks so short that the studs lack support.

For the greatest bracing strength, you can install the diagonal bracing from the bottom of the third or fourth stud to the corner post at a point 4 feet above the soleplate. Do this along both walls. Then start the next diagonal bracing 6 inches above the end of the first bracing and to the third or fourth stud. If you do this in both directions at each corner you will have a remarkably strong corner that will not sag, sway, warp, or in any way settle to an out-of-square position. (See Fig. 6-6.)

If there is a wide or tall window opening in the wall frame, it is good to extend the bracing so that it crosses at the corner of the window and extends to one cripple stud across the bottom of the window. (See Fig. 6-7.)

Do the same at the bottom of windows. As mentioned earlier, the window often creates the widest span of otherwise unsupported area in the entire room. Windows, therefore, are potential weak spots. In older houses the doors work well for decades, but in a large number of cases, the windows provide difficulties.

When all the bracing is done, you are ready to nail in the cap plates. This is a very important step and one that must be done with care and considerable planning.

Chapter 7 is devoted largely to the top plates, their importance, and the problems associated with the steps, and to the top caps, which are crucial to the interior walls.

7

Top Plates and Caps

THE EXTERIOR WALLS ARE IN PLACE AT THIS TIME. YOU ONLY NEED
to install the double top plate and cap the walls. For top plates, you need the longest and
truest 2 × 4s you have. These are nailed along the first plate that you nailed earlier over
the studding.

INSTALLING TOP PLATES AND CAPS

The double top plate strengthens the wall by adding 2 more inches of framing to support
the roof, but you actually do far more. The single plate connects the studs, and you perhaps had to join the plate sections, since you probably did not have a single 2 × 4 long
enough to reach from corner to corner of the wall. The joining occurred midway across
the top of a stud, which means that the weight supported by that particular stud is really
supported by less than 1 inch of 2 × 4. The milled stud, which is sold as a 2 × 4, is
actually considerably smaller. The stud was a true 2 inches by 4 inches before it was finished. All irregularities were trimmed, planed, or sanded off, leaving what amounts to
a 1⅝ by 3⅝ timber. The end of the top plate that is joined atop the stud, then, is resting
on one-half of the stud top.

The double top plate is added, therefore, to add strength and to stagger the joints.
If one joint occurs at the top of the fourth stud, the joint of the double plate will end three
or four studs away. In this fashion all studs and spaces between studs will be supported
by one full timber and at least a portion of another. Your wall frame will be considerably
stronger and will support roof weight without difficulty.

This staggering of timber ends is called *lap nailing*. You should use 16d nails for
this purpose.

The next step is top capping the wall frames. Here you must know where all wall
partitions will join the exterior walls before you can nail in the top caps.

If you are working from detailed plans, you already know where the partition walls will be. A large number of do-it-yourselfers, however, work with only a general idea of house plans. Professional builders might also occasionally work with inadequate plans. One builder, described by his customers and competitors alike as the best in the area, once admitted to us that he hadn't worked from architectural plans in ages. He pulled a dirty and wrinkled scrap of paper from his pocket and asked us to study his house plans. He said that a huge percentage of his plans were drawn by amateurs on the backs of grocery bags.

The reasoning is clear. The architect will charge as his fee a percentage of the cost of building the house. A fee of only 2 percent amounts to $1600 on an $80,000 house. If the fee is 5 percent, the fee is $5000 for the same house. Many, if not most, of the do-it-yourself workers want to save money, and they reason that detailed plans are not needed in order to add a room to an existing house or to build a rather modest residential structure.

If you are working from plans on the back of a grocery bag, or if you are using the plan-as-you-go approach, you need to stop at this point. Before the top capping begins, determine without question where your room partition walls will be erected, because the gaps in the top capping allow the walls to be tied together effectively.

If you already know the location of these partitions, then start capping. Use 2-×-4 timbers that are long, true, and free of flaws. Use lap nailing so joints will not end at the same place. Be sure to leave room at one of each corner wall for a 2 × 4 to be nailed in place. Do the same thing anywhere a wall of any sort will butt into the exterior walls. (See Fig. 7-1.)

When this is done, you are ready to start assembling the interior wall frames. If you decide later that the wall you had planned to join the exterior wall at a specific place simply will not work properly, you can make the necessary changes by taking up the top capping and reworking the plans.

If you are building an entire house—or enough of one that you will have several interior walls—here's how you can start to construct the walls.

First, determine where the wall will be erected and use your chalkline to mark the location. Double check on the measurements before you go too far. Mark the beginning of the wall line on the soleplate, and do the same thing on the opposite soleplate for the end of the line. Then measure from the exterior wall to both points, to be sure that the room will be square. You can also make a diagonal measurement for double certainty. When you are satisfied, chalk the line and start to lay out the lumber for the partition wall.

As you did before, lay out a soleplate and stand it on edge with the facing toward you. Beside it and parallel to it, lay out the top plate. Mark with a line and X where studs will be nailed in, the X on the inside of the soleplate and outside of the top plate, if you plan to raise the wall as a unit rather than build it piece by piece.

When this is done, move the top plate 8 feet away and lay out the common studs, partition studs, cripples, if any, and the corner posts or partition posts. If there is to be a doorway in the wall, build the header and lay it in position. Note that the soleplate extends across the doorway. When you are nailing down the soleplate, however, do not put nails between the door frame studs. This portion of the soleplate will be cut away later, and you don't want to have to pull nails.

Fig. 7-1. The first length of top capping has been installed at the top of the window shown above. The next length will be butted against the end of the first timber. If you plan to have a partition wall meet the existing wall, leave a space the width of a 2 × 4 between timbers.

When you have nailed in corner posts or partition posts and all studding, as well as headers and rough doorways, you can lift the wall section into place. Now you must be sure that the end of the wall butts accurately into the partition studs at either end. Check to see that the soleplate is resting with the inside edge barely touching the chalkline.

If the soleplate is not located properly, use a hammer to bang the soleplate until the wall frame moves into the right location. Nail the partition wall frame to the partition studs or posts, and nail the soleplate, except for the portion inside the door frame. When this is completed, you can nail in the top capping and tie the walls together permanently. (See Fig. 7-2.)

If you are building the wall piece by piece, do so the same way you did the exterior walls.

Before you leave this particular job, use the level and square again to be sure that your wall is still plumb, vertical, and square.

INSTALLING WINDOW AND DOOR FRAMES

At this point you can start to rough in window and door openings. Before the wall units went up you had only left a wide opening where the windows and doors would be installed. Now it is time to nail in all the trimmers and other timbers.

First of all, the window frame is built in accordance with the size of the window, but the frame itself will not be the exact size of the window itself. If this sounds confusing, remember that there must be some clearance in order for the window to move freely, and you must then enclose the window so that it is firmly positioned.

If you have not already done so, you should buy standard-sized windows, except in very unusual circumstances. If you do buy standard-sized windows you can simply position the window in the rough opening and save a great deal of time and labor.

Consider the fact that ready-made draperies are always made for standard sized windows. If you have an odd-sized window, you will either have trouble in finding the draperies or you will have to pay considerably more for them when you do find them. Be sure that your windowpane size is also standard. If it is not, you will again encounter difficulties.

When you start to frame the window, first cut the trimmer studs. These will be installed by nailing them to the inside of the regular or normal studs that mark the rough window opening. The trimmer studs will be cut into two pieces.

To easily combine window framing and window header installation, nail trimmer studs to the normal studs first, and then simply slip the header into position. The length of one trimmer piece will reach from the header to a point where the rough sill will be located. The other piece will reach from the bottom of the rough sill to the soleplate. In other words, leave enough space so that you can nail in a 2 × 4 for the rough sill.

Before you nail the trimmer in place, be sure that the ends are cut square. Otherwise, the header will not seat firmly and securely, and the rough sill will not fit snugly into its slot.

Nail in the two pieces of the trimmer stud. Measure carefully, however, to be certain that you have marked the spot for the rough sill accurately.

Now position the header so that it sits squarely on the top ends of the trimmer studs. They offer excellent support for the header, which cannot sag as long as the trimmer studs

Fig. 7-2. In this photo you can see how the beginning of the top capping (at the top left of the photo) is installed. The short piece that butts against the installed piece will tie the two wall sections together.

maintain their strength. For this reason, select trimmer studs that are sound and as free of flaws as possible.

If you have already nailed in the header, you can simply cut the trimmer studs, fit them under the header, and nail them in place with 16d nails. Use one nail every 15 inches or so for the trimmers.

Now your rough window opening extends the full length of the height of the wall, except for the header. Your next step is to install the rough sill.

If you have any reservations about your expertise in leaving the slot for the rough sill, take the easy way out. Measure the distance from the soleplate to the bottom of the rough sill and cut two lengths of trimmer studs. Nail these in place. Then set the rough sill in place atop the ends of the trimmer sections you have just installed.

You can nail in the rough sill, which is a length of 2 × 4 that will reach from stud to stud and span the rough window opening, in one of two ways. One simple and easy way is to nail through the top of the rough sill and into the end of the lower trimmer studs.

You can also nail through the common stud and into the end of the rough sill. The argument for nailing in this fashion is that you strengthen the window opening by guaranteeing that the studs will not sway or bend later and create a loose window.

If you wish, you can nail in the rough sill in both ways. Nail one 16d nail through the rough sill into the lower trimmer stud, and two through the regular stud into the end of the rough sill. Do this on both ends and the rough sill will be stabilized.

Now measure the distance between the top of the rough sill and the bottom of the header and cut two sections of trimmer studs the exact length. Nail these in place with 16d nails spaced 12 to 15 inches apart.

Measure to see that the rough opening is exactly the same distance across at top and bottom. Then measure diagonally and compare measurements again. You should get the same reading at top and bottom, and the two diagonal readings should be exactly the same.

That completes the rough window framing. You are now ready for the rough door frames.

One of the first things you will notice about the rough door framing is that it is very similar to the window framing in the opening steps. One difference is that the trimmer studs will not be installed in two pieces, since there will be no rough sill to install.

If your header is assembled and ready to install, measure it from top to bottom. Then measure and cut two trimmer studs that are the length of a normal stud minus the depth of the header. If the header is 6 inches, again cut the trimmer studs six inches shorter than normal studs.

As you did on the window frame, check to be sure that the tops of the trimmer studs are cut squarely so the header will be seated evenly and firmly in place.

Nail the trimmer studs to the regular studs by using 16d nails spaced 15 inches apart from top to bottom. It is a good idea to stagger the nails rather than nail in a row, to evenly anchor the trimmer. (See Fig. 7-3.)

Rough door openings, as a rule, are 2½ inches wider than the door itself. So if your door is 32 inches wide, you will need an opening that is 34½ inches wide. This opening is required for the addition of the door jambs, any wedging that is needed, and clearance for the door to swing.

You also need to allow space vertically. If your door is 6′8″ high, you will need to allow 6′10½″ for sufficient clearance, plus the addition of the head jambs.

Fig. 7-3. Install trimmer studs at the sides of each doorway. An easy way to install them is to stand them beside the regular stud and sink nails through the top of the top plate into the end of the studs. You can then sink nails from the outside edge of the trimmer into the side of the installed stud.

When you have nailed in the trimmer studs and the header is in place, you should get out your measuring tape and make another set of measurements. Measure the distance across the doorway in three places: top, bottom, and middle. You should get the same reading in each place. Next, measure from the bottom left corner to the top right and record the measurement. Then measure from the bottom right to the top left and compare the two readings. If there is a serious discrepancy, corrections should be made at this time.

Much later, when you are ready to hang the doors in your room or house, you will have to make several decisions. What you have done thus far will determine, to a large extent, what you must do later. You will need to decide, first, whether you are going to complete the door framing and finishing yourself, or whether you will buy a door that is already framed. You can save several dollars by doing the framing yourself. If you plan to buy the preframed door, you should check to see what the dimensions of the frame and door assembly are, and, if you need to, make needed adjustments before you complete the rough framing of the door.

You will need to decide, too, which side of the door will have the hinges and which will have the lock. Even though the room is not completely framed, you can start to envision any problems that will surface later. Will the swinging of the door block all or part of a window? Will it block other doorways, such as those of a closet, pantry, or cabinet? Will the swing area of the door present problems with furniture or appliances?

It is easy to change your mind and make modifications at this stage of construction. Later, it might not only be difficult, but time-consuming and expensive.

Here is one example of preliminary testing steps you can make. If you have a door available, stand it in the doorway and measure the distance between the top and the header. Now add the thickness of the materials that must go between the door and header: the head jamb, flooring, and clearance space. The head jamb will be $\frac{3}{4}$ inch thick, and the flooring will also be $\frac{3}{4}$ inch. You will need $\frac{1}{16}$-inch clearance in addition, so you should have at least $1\frac{1}{16}$ inches clearance, in addition to the threshold, which will be another $\frac{3}{4}$ inch—a total of $2\frac{5}{16}$ inches.

A good check to make later, after the door has been fully framed and the door is ready to hang, is to tape two 4d finishing nails to the top of the door and then move the door into position. Tape the nails so that the heads protrude over the edge of the door and the nail shanks cross the top of the door. Now move the door into position to see if it fits. If it does, you have the door exactly right, because the shank of a 4d nail is $\frac{1}{16}$ inch, which is the standard clearance for doors. The fit should include the threshold, if one is to be installed, and all parts of the finished door frame in place.

You can also test the door for proper clearance on the sides by using the same taped nail test, since side clearance is the same for sides as it is for the top. But remember to tape nails to both sides and at the top and the bottom for best results.

Store your windows and doors safely until you are ready to hang them. If you can leave them at the supply house until you need them, do so. If you leave them on-site, you are inviting theft and vandalism, unless you are simply adding a room or two and live on the property. There are several reasons for buying the windows and doors early. Perhaps they were on sale at a great savings and you could profit by buying early. Sometimes the items can be bought at a clearance sale and you can not afford to wait. Maybe you could get a construction price on all of your lumber and other materials if you bought in large quantities at one time. You might have temporary access to a truck and could save delivery fees by hauling the doors and windows yourself. Under these circumstances, you might be justified in buying and storing the windows and doors before you need them.

Among the reasons for waiting to buy these units in addition to the possibilities of theft and vandalism, are the damage resulting from weather and accidents, improper storing, and exposure to cold and heat. Your window panes could be broken. Doors can be warped and scratched, knobs damaged, and edges, which are extremely susceptible to weathering, swollen or marred by carelessness or wear and tear.

Here are a few tips for storing these expensive materials, if you purchase them in advance. Do not stand doors on edge; instead, store them flat on the floor. Do not store them on a pair of 2-×-4 studs, since the doors will tend to sag in the middle, and the studs perhaps will leave their imprint on the surface of the doors. Do not stack more than half a dozen doors in one space. This is especially true if knobs and hinges are already on the doors, in which case damage is almost a certainty.

Be sure that you store the doors and windows in a sheltered, well-ventilated place that is dry at all times. It has been said that a foggy morning is sufficient to cause a stored door to swell or warp, and whereas the claim might be somewhat overstated, if moisture is allowed to seep between the stored doors and remain there for several days, you are inviting trouble.

Cover the doors and windows with some material that will protect them from moisture as well as from direct sunlight and wind. If you must leave your doors stacked for more than a few days, wrap edges with plastic and cover the stack with canvas or a similar material.

Do not handle the materials any more than is necessary. In addition to the danger of breakage or scratching, there is the problem of oil from human hands and the dirt and grime from your work that can stain or soil surfaces.

Store the doors and windows away from the busiest work areas and out of the traffic flow. You'd be surprised how often such stored materials either are stepped on or have tools or lumber dropped on them.

If you must move doors, be sure to carry rather than drag them. Dragging damages the tops or bottoms, and you may drag the door over nailheads, tools, or pieces of lumber.

If a door is soiled unavoidably, stop work and clean it at once. If the grime is permitted to stay on the surface, it will be ground into the surface by subsequent handling and become very hard to remove. Keep all stored doors and windows away from wet plaster, mortar, or concrete.

When your wall framing is completed, you may wish to install doors and windows immediately to keep them out of harm's way and to secure the building site against uninvited visitors. The installation of doors and windows can be a reasonably difficult operation, unless you move carefully and check and double-check your work. If, however, you proceed with caution, the work can be accomplished with few if any real problems.

Among your major tasks will be, in addition to the actual framing of the doorway, the installation of hinges and locks, and the blocking or wedging for perfect door fittings. Installing deadbolt locks, which is highly advised for all exterior doors, can also be difficult, unless you have the proper tools and some knowledge of correct procedures. Striker plates can also cause damage if you do not work carefully.

Chapter 8: Installing Doors, takes you from the rough door opening through the entire process of putting up a door.

Installing Doors

DOORS OCCUPY A SMALL PORTION OF A ROOM, BUT THAT PORTION IS among the most important areas of the house. A poorly fitted door might stick, squeak, drag, or scrape; it might admit moisture, hot or cold air, and insect pests. You can pack walls and ceilings with insulation, but one poorly fitted door can negate a great deal of your work. A poorly fitted door can cause locks and latches to malfunction. Such a door might also damage carpet or hardwood floors.

FINISHING DOOR OPENINGS

When the rough door opening is framed and you are ready to hang the door, you can make one last check to see that you have a perfect rectangle. Then your next big task is to see that the headers are level before you nail in the jambs and blocks.

You will need to buy the jamb units, unless you are equipped to cut *rabbets*. These are grooves cut in the face of a board so that another board or other material may be fitted into them. The rabbet is difficult to cut, so you will be better off buying the jambs. If you bought the entire door assembly, you probably bought the complete door, frame, and casing assembly, so you don't have to worry about the rabbet work.

Nail in the *jambs*, which are the linings of the door openings. Casings and door stops are nailed to jambs, and the door is hung from them. Jambs for interior doorways are usually 1-inch boards with the rabbet cut in them (with the finished thickness usually ¾ inch), and exterior jambs are 1½ inches thick.

We suggested that you level the header before you do anything else. The reason is that if the header is not level, then the head jamb installed below it may not be level either. If you need to, you can use wedges between the header and head jamb to get a perfect level for the jamb.

You also need to check the flooring inside the doorway to see if it is level. If it is not, one of your side jambs will be higher than the other, and the result will be an unevenly hung door.

When you cut the head jamb, square both ends so that they can fit snugly into the rabbet or dado cut. Cut the head jamb long enough that it will reach across the door opening plus the depth of both dadoes or rabbets. (See Fig. 8-1.)

When you are cutting the side jambs, measure from the bottom edge of the rabbet or dado downward. Your measurement should include the length of the door, plus clearance needed under the dado, generally $\frac{3}{16}$ of an inch. Cut the jamb accordingly. Vary the length of the two side jambs only if one side of the doorway is higher than the other. For instance, if the left side is $\frac{1}{2}$-inch higher than the right, then the left side jamb should be $\frac{1}{2}$-inch longer, so that both jambs reach the floor and rest snugly there, with a good fit.

When the jambs are cut, assemble them into a door frame by inserting the head jamb into the rabbets or dadoes cut into the side jambs near the top. The parts of the side jambs that extend beyond the dadoes are called *horns*. With the head jamb inserted in the dadoes of the side jambs, nail the head jamb into place by driving an 8d nail through the side jamb and into the end of the head jamb.

Now cut small blocks of wood the thickness that your finished floor will be. These blocks should be about five inches long and about the same width. Set these blocks against the sides of the doorway and then set the jamb assembly upon the blocks in the doorway. Because the blocks are cut the same thickness as the floor will be, the jambs resting upon the blocks should be exactly right for your door.

Next, fit wedges or shingles behind the jambs on either side of the doorway. Use shingles that are just thick enough to wedge the jamb into a perfectly plumb position.

Use the longest level you have—one five or six feet long if it is available—for the jamb's reading for a plumb position. Hold the level tightly against the jamb. If there is any irregularity in the jamb that might cause an incorrect reading, remove it.

If you have to use a shorter level, take the reading at a series of locations up and down the jamb. When you are satisfied that the jamb is plumb, nail it securely in place. Be sure to nail through the jamb, wedge, shingle, or block; and into the stud.

Next, straighten the other side in the same way. When it is plumb, check the head jamb again for level, and if it is still correct, nail the other side jamb in place.

FITTING A DOOR

First determine which side of the door will be hinged. Then mark the proper points on the side jamb or, as it is sometimes called, the *stile*, where the hinges will be located. For interior doors, the hinges usually are placed $4\frac{1}{2}$ inches from the head jamb and $11\frac{1}{2}$ inches from the bottom jamb. That is, the top of the *hinge pin* (the loose pin, usually 4 inches long, that slips into the hinge openings) should be $4\frac{1}{2}$ inches from the bottom of the head jamb and the bottom of the pin receptacle should be $11\frac{1}{2}$ inches from the floor. Measure carefully and mark the proper points. If you prefer to mark the exact location of the hinge rather than the pin, the top of the hinge should be exactly $5\frac{1}{2}$ inches below the head jamb.

If you need to make any adjustments on the door opening, set the door aside, measure and change, then try the door again. Try the door in the opening often, because you want the best possible fit. The proper clearance on either side of the door and at the top is $\frac{1}{16}$ inch. Clearance at the bottom depends whether you plan to install a threshold or not and whether you will use hardwood flooring, carpet, or especially thick carpeting. For normal

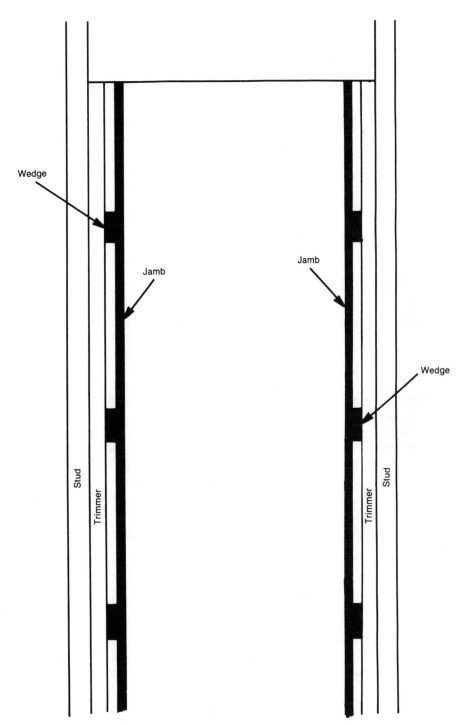

Fig. 8-1. The side jambs are installed against the trimmer studs. If the jambs are not perfectly vertical, you can nail in small wedges or shims to provide a true vertical positioning.

carpeting, the clearance, without a threshold, should be ½ inch. For thicker carpeting the clearance will have to be greater, depending upon the actual thickness of the carpet pile.

The typical door hinge is a *loose-pin butt mortise*. This means, in addition to the feature of the loose pin that can be extracted easily, that each hinge leaf will fit in a mortised fashion in the door jamb and in the edge of the door itself. The leaf of each hinge will mesh with the other in a tight and neat fit.

Interior doors usually have only two hinges. Exterior doors, which are heavier, might have three hinges, and these hinges might be larger to support the extra weight. If your interior doors are thicker or heavier than the typical door, you might add extra hinges or use larger ones. If the door is from 1⅛ inches to 1⅜ inches thick, and 32 inches wide, you will need at least a 3½-inch hinge. That is, the hinge from top to bottom, not counting the loose pin, is 3½ inches in height, and the width is about 3 inches. Each leaf is about ⅛ inch thickness. Doors up to 1⅞ inches thick and 32 inches wide need 4-inch hinges. Doors that are 37- to 43-inches wide need 5-inch hinges. If the door is especially heavy, it will need heavier hinges.

CONSTRUCTING AND USING A DOOR JACK

If you are going to hang your own doors, you will benefit from a simple device called a *door jack*. Although it is not a jack at all in the usual sense of the word, this simple instrument will help you set hinges, cut bevels, or do the other work required for hanging or finishing doors. The door jack holds the door erect while you work on it, without causing damage to the door.

You might have to bevel one edge of the door in order for it to close without binding. If so, do not prop the door against a wall or sawhorses. It is very awkward to try to hold the door while you perform the work, and you might not always have someone around to hold it for you.

Even if you have a helper, it is not economical to tie up his services when a simple-to-make and inexpensive device such as the door jack is so readily available.

To construct the door jack, you will need 1-×-6 or 1-×-5 boards, one of which must be at least 8 feet long. If you wish, you might use 2-×-6 lumber for the 8-foot base piece and 1-×-5 or 1-×-6 lumber for the rest of the jack.

Lay the 8-foot section on the floor and nail a 1-×-2 or 2-×-2 brace at the very end of the base piece.

Next cut a 3-foot length of 1 × 5 or 1 × 6 and square one end. At the other end cut a V or notch at least 5 inches deep, and wide enough for a door edge to fit into. This portion of the door jack is the *jaw* piece.

The next piece of 2-×-4 stock is 3 feet long. On each end there is a 5-inch block of 2-×-4 stock nailed to the bottom of the 2-×-4. Put this cross piece aside until it is needed. The next piece is a brace that is made of 1-×-5 or 1-×-6 stock, cut 20 inches long.

Temporarily nail the base piece (with the brace or block nailed to the end) to the floor. Then nail the 20-inch-long brace to the cross piece so that the brace sticks straight up and the blocks on the ends of the cross piece are toward the floor. The upright brace piece should be parallel to the cross piece. It is centered on the 2-×-4.

Now place the jaw piece so that the square-cut end is against the end brace of the base piece. The jaw end is raised so that it can rest atop the upright brace. Then nail the jaw piece to the upright brace. At this time you should have an assembly that includes the 8-foot-long base piece, which is nailed to the floor temporarily. A jaw piece rests against the brace or block at the end of the base piece. It is nailed to the end of the upright brace, which in turn is nailed to the cross piece. The cross piece is positioned across the base piece. (See Fig. 8-2.)

If you wish, you can wrap the jaws with scrap cloth or similar padding. This will prevent the jaws from scraping or otherwise damaging the end of the door.

To use the door jack, stand the door on edge so that one end of the edge rests upon the cross piece and the top edge is set inside the jaws of the jawpiece. The rest of the door rests on the base piece.

This door-jack assembly will hold the door in an upright position while you plane, bevel, or sand it. You don't have to worry that the door will slide and be damaged.

SETTING HINGES

A good time to try out the door jack is for the process of setting the hinges for the door. Start by laying the hinge at the point where you want it installed and then making a pencil outline of the hinge. Lay the hinge so that the top is about 4½ inches from the top of the door and so that the edge of the hinge lies perfectly straight across the door edge except for a tiny strip about ¼ inch wide. In other words, you don't want the hinge to extend completely across the door edge, because when you chisel out the gain or opening for the hinge, you don't want the hinge to show when the door is closed.

When the hinge outline is completed, move to the bottom of the door edge and mark the placement for the second hinge. Lay off the position as you did before.

You will need a wood chisel or other sharp-edged tool that you can drive with a hammer into the wood one-eighth of an inch all around the outline. The point of the chisel will have one straight edge and one beveled edge. Place the point, the bevel edge, inside the outline, so that the blade is on the pencil mark. Hit the chisel gently with a small hammer until the blade has been driven ¼ inch into the wood.

Using the same procedure, follow the outline of the hinge all around the perimeter. When the outline has been cut, move in about ½ inch and place the blade of the chisel against the wood. Tilt the chisel to a 45-degree angle. Hit the chisel gently until you chip out the wood up to the first cut.

Move the blade over slightly and tap the chisel again, and little by little you will gouge out the wood next to the pencil outline. Do not hit the chisel hard, because excessive force will cause the blade to split beyond the outline and severely damage the door edge.

Now chisel out the rest of the wood inside the outline until you have a perfect rectangle that is uniformly smooth both along the edges and on the interior surface. From time to time lay the hinge in the gain (or recess) to test for a fit. When the hinge fits perfectly and its edges are exactly flush with the edges of the door, your job here is completed.

Go to the next outline and do the same. These cut-outs are called the *hinge setbacks*.

At this point, lay out hinge outlines on the side of the side jamb. Measure carefully so that when the hinges are mounted on both door and jamb, they will match perfectly and the door will be in its desired location.

On both jamb and door edge, the hinge should protrude over the edge slightly, about ¼ inch. This space is needed for the door to swing easily and close properly.

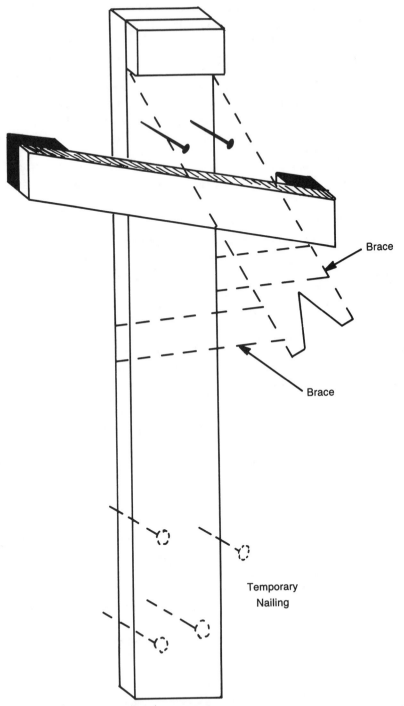

Brace

Brace

Temporary
Nailing

Fig. 8-2. The door jack is used by placing the door top in the V of the top timber. The V will hold the door erect and steady while you work on it.

Chisel out the setback in the jamb as you did in the door. Mark the outline, cut along the line, then chisel out the inside of the rectangle. If you have any part of the setback that is higher than the rest, your hinge will not seat properly. To get rid of raised places, lay the chisel over until it is almost flat, then tap the handle gently with your small hammer.

You are now ready to install the hinges. To do so, take the loose pin from the hinge and separate the leaves. Check to see that the loose pin is pointed in the right direction. You must have it positioned so that you can pull upwards on the pin to remove it from the hinge. If it faces the other direction, it can fall from the hinge and the door could actually fall from its mount.

Place the leaf of the hinge in the setback you chiseled earlier and position it in the best location. Remember that you want it completely flush with the edge of the door. While the leaf is in position, use a pencil to make a small circle inside the screwholes.

After marking all three holes (or four, if you have heavy-duty hinges), remove the leaf and place the point of a nail in the center of each hole. Tap it lightly with a hammer until the point has been driven into the wood half an inch. You will find that the small hole made by the nail will permit you to start the screws much easier.

You should mount all four of the hinge leaves—two in the edge of the door and two on the jamb or stile of the doorway—so that all that remains is to position the door and insert the loose pins.

One of the easiest ways to mount the door is to lay a wood strip about ¼ inch thick and 12 inches long across the door opening near the stile or jamb. Set the door on the strip and angle it to a position of about halfway open but upright.

If you put the door into the closed position, you will have difficulty both in holding the door and in seeing the position of the door hinges to those of the jamb. If you hold it in the half-open position you can, while standing, see all critical locations and at the same time be in position to move the door as you need to do. While watching the upper hinges, gently ease the door into the exact position so that the hinge leaves will slide together and you can hold the door with one hand while with the other you can drop the loose pin into the hole created for it.

Before you start this operation, it is wise to have the loose pin in a pocket or other accessible place and your hammer within easy reach, because the loose pin may be tight or the leaves of the hinge may not be aligned perfectly and the pin will not drop. If you can start the pin by hand, you can use the hammer to tap it very gently into position.

Do not batter the head of the loose pin by trying to force it into place. If it does not respond to a gentle tap, you need to reposition the hinge leaves by gently jockeying the door back and forth while pushing toward the hinge at the same time. If you can get the loose pin partially started, so that the door will stay in place if you turn it loose, you can lay a strip of wood on top of the loose pin and tap the wood, instead of the head of the pin, with the hammer.

When the first loose pin is fully seated, you can hold the door by the outside edge and maneuver it so that the bottom hinge leaves match. Push the door hard enough that the hinge parts mesh and then insert the loose pin as you did before.

When both loose pins are in place, move the door to and fro so that you can see whether it will swing freely. If you find that the door binds, you will need to push it gently forward until you can see where the problem occurs. If the door will not close fully, you may have to take it off its hinges and place it again in the door jack and, using a plane, bevel ¼ inch off the inner edge of the door.

If the door scrapes at any point, you may have to reposition the jambs. Be sure you move the wood strip on the floor before you try to open and close the door.

INSTALLING LOCKS

Once exterior doors are installed, you should proceed with the installation of locks so that the house or room will be more nearly vandal-proof. The easy way to handle the lock problem is the one most people choose: calling in the professional.

You can install a lock yourself, however, with very few tools and only a few minutes of time. The first installation takes more time than subsequent ones, so don't be discouraged if your initial effort proves to be disheartening.

First, there are obvious problems to avoid: any time you use a chisel, there is always the possibility of splintered wood, and you can seriously deface the door by overly energetic use of the chisel and hammer. It seems that most people make their worst mistakes when they are tired and nearly finished with the job, so try to remain as patient as possible. You may find, though, that in spite of all of your efforts, there is still some frustration. In such an event, leave the job and concentrate on something else that requires less meticulous effort. If this sounds as if lock installation can be vexing, it's because it can very well be.

There is only one major problem, and that is the one of making the hole or holes in the door. Depending upon the style of lock you choose, you may have the option of making one 2-inch hole or two 1-inch holes.

The term *make a hole* is used advisedly, because you have the options of boring holes with a brace and bit or auger, or drilling them with an electric drill. A third possibility is that of boring a small hole, then using a keyhole saw or hacksaw blade to cut a circle in the wood.

The easiest method is to use an electric drill with one of the special hole-cutting attachments. You can buy the set of hole-cutting tools for less than ten dollars, and it might well be worth the money, especially if you will be installing several locks or door-knob assemblies.

The drills described here are extremely simple to operate, and they can be attached to virtually all electric drills. The tool consists of a circular base with grooves to allow the use of several sizes of hole-cutting attachments. The basic drill has a short ¼-inch drill that extends beyond the hole-cutter. The hole-cutter bit is locked into place so that there can be no variation in the size or shape of the hole.

As you use the drill, the ¼-inch bit can be placed in the marked center of the hole, and it will sink quickly so that the hole-cutter will come into almost instant contact with the wood. This latter part of the drill consists of a circular bit or blade with finely serrated edge. As soon as the circular blade reaches the wood, you will see a perfect circle appear in the wood surface.

As you continue to work the blade will sink slowly but deeper and deeper into the wood. Do not try to force the blade to cut faster than its normal rate. To do so could result in burnout of the drill or damage to the blade or door.

Your major problem will occur when the blade emerges from the other side of the door. No matter how carefully you bore or drill, there is little way to avoid splintering the opposite side of the wood, unless you use the following procedure.

First, before you drill anything, you need to decide where the lock assembly should be located, and you need to take care that you drill straight through the door. If you bore or drill at an angle, you can produce serious problems.

To mark the lock location, open the door to a point that you can work comfortably. If you have small wedges or wedge-shaped scraps of wood, you can stabilize the door by putting a wedge under each side to hold it steady while you work.

The usual location of locks is three feet above the floor, so you can measure up 36 inches and mark the edge of the door at the proper place. Use your square lined up with the edge of the door and mark the lock location in line with the squared line. Now do the same thing on the door post, stile, or jamb where the lock bolt will enter.

The lock assembly comes with a template or pattern you can use for marking all locations. Fold the template along the dotted line and then hold the template so that the dotted line is on the edge of the door.

With the template in position, one part of it shows where the 2-inch hole on the face of the door must be. The other half of the template indicates where you should drill for the lock bolt. The center of the two-inch hole will be 2½ inches from the edge of the door.

With the template still in position, you can mark lightly with a pencil the top and bottom locations of the template. Then you can remove it, turn it, and reposition it according to the marks you just made. Mark the 2½-inch location on the other side of the door. Use great care in making the second mark; it must line up exactly with the first marks.

The purpose of the second mark is so that you can start to drill on one side, and when you are halfway through stop and then drill again on the opposite side. When your second hole meets the first one, you can remove the drill and the section of 2-inch wood, and there will be no splintering of the door face.

Go next to the edge of the door, to the mark you made at the exact center, and choose a cutting blade or drill that is slightly larger than the bolt for your lock. Drill straight into the edge of the door, and be sure that the hole is a little deeper than the length of the bolt.

You need to outline the latch-bolt mounting plate now. When you have done so, chisel out the wood so that the plate can be mounted flush with the edge of the door.

You are ready at this time to install the lock assembly. The assembly including the knob for one side slides through the 2-inch hole, and the opposite knob will attach to the other side. Long bolts enter from both sides of the door so that it will be impossible for an aspiring burglar or vandal to remove the lock by removing the bolts from only one side.

The bolt is installed through the hole in the edge of the door, and the lock is installed, except for the strike plate for the jamb or stile.

Again, you must cut a gain and drill a hole large enough for the bolt to slip through. Before you cut the gain, position the strike plate and outline it, then chisel out the wood inside so that the strike plate fits flush against the surface of the jamb or stile.

One way to guarantee a perfect fit is to apply a bit of chalk to the end of the bolt before you do any cutting or drilling. Then close the door and lock it. The bolt will butt into the jamb or stile and leave a faint chalk outline. Open the door and mark the chalked spot and outline the strike plate. In this fashion you will not have to worry about a perfect lock fit.

If you want double protection against vandals, you can install a dead bolt lock, which cannot be opened with any of the standard devices such as fingernail files, credit cards, tiny screwdrivers, and the like. You will follow the same approach as you did for the regular lock assembly, except that it will be installed slightly below the doorknob assembly. If you prefer, it can be put in place at any point on the door that you wish.

FINISHING THE DOOR INSTALLATION

There is little more to do at this time on the door. You can mark the location for the door stop trim and then cut and nail it in place. One easy way to install this molding is to close the door to its normal position and then place the molding against the door so that there is only a $\frac{1}{16}$-inch clearance. Nail in the first piece of molding or doorstop only after you have mitered it at the top. The bottom should be cut square so that it will fit evenly against the floor. Miter the second piece, and with the door still closed you can position it in the same way, then nail it in.

You will need to cut a third piece for the top and miter both ends. Slip it in place and nail it securely. You can use finishing nails for the entire process.

Now it is time to nail in the casing or facing. Cut 1-×-4 boards long enough to reach from the floor to the head jamb. You will need four of these—two for either side of the doorway. Nail these in place and then cut lengths to fit over the casing boards you have just installed. You will also need two of these—one for either side of the doorway.

Be sure that all cuts are perfectly square so that the fit will be good at all points. Use finishing nails to install the door facing.

If you decide to add molding now, you will need two pieces of molding the length of the door height plus the casing above the door. In other words, the length is the height of the door plus $3\frac{1}{2}$ inches. You will need one piece the length of the top casing. Miter cut both of the side pieces and nail them in place, using small finishing nails. Then cut the top piece and install it.

Do the same on both sides of the door and your work here is done for the time being, unless you decide you would like to try to make your own doors.

First cut side pieces the length of the door and 4 inches wide. Then cut the top and bottom pieces, also 4 inches wide. Saw a groove in the ends of both long pieces and then saw out a tongue on the ends of the short pieces so the tongues will fit in the grooves. Saw kerfs or grooves on the inside edges of all the outside pieces.

Cut thin panels ($\frac{1}{4}$ inch thick) that will fit inside the kerfs you sawed. You will need thicker panels, also with kerfs sawed on both edges, for the center for the door. Fit the thin panels inside the kerfs. When you either nail or otherwise fasten the sections securely, you have a door that can be made in an hour and at a cost of less than one dollar—provided you saw your own lumber, as we did, with a chain saw. Otherwise the door will cost you about four dollars. (See Figs. 8-3 and 8-4.)

You are now ready to install windows, if there are to be windows in the wall you are framing.

Fig. 8-3. After you cut the side and top pieces for the door, and after you have cut a groove or kerf along the inside edges of these pieces, you can install the thin sheets of wood on the sides. Then the center piece, also kerfed, can be tapped into its final position. Use a block of wood rather than hit the top of the panel with the hammer.

Fig. 8-4. The finished door, with all cuts made with a chain saw, cost less than 50 cents to make. The door can be left with its natural grain showing, or you can stain or paint it.

9

Installing Windows

IF YOU HAVE ALREADY INSTALLED WINDOW FRAMES, YOU CAN proceed directly to the installation of the windows. If you have not yet installed frames, now is the time to start. Refer to the previous chapter for general help in installation processes.

You soon will notice that there is a strong similarity between framing of doors and windows. Both have headers, and both have trimmer studs and facing. Both have cripple studs, but those of doors are on top, while the window frame includes cripples both above and below.

PARTS OF A WINDOW

To install a window, you need to know a few basic terms, including upper and lower stiles, top rail, bottom rail, horizontal bar, upper meeting rail, muntin, and sash. Some of these terms have been used previously.

You must have, first, a window frame that is constructed according to the dimensions of the window to be installed. You must also include a clearance allowance in order for the window to have room to slide up and down without sticking or binding.

The *sash* of a window is, technically, the part of the window that holds the panes in place. Here, and elsewhere, however, the term includes the entire window assembly: frame, panes, and the rest of the entire assembly.

The *upper stile* of a window is the frame unit that extends the entire length of the window on both sides and acts as a frame for the top rail and upper meeting rail. In other words, the typical window assembly includes two sections of panes, both of which slide inside the tracks or rails. Each section contains panes of glass, and the side framing of each sash is a stile.

The *lower stile* is precisely the same thing, except that here we refer to the bottom half rather than the top half of the window unit.

The *top rail* is the framing that bounds the top panes of the window and is installed between the two top stiles.

The *bottom rail* of the upper half of the window is called the *upper meeting rail*, since it meets and overlaps the top of the bottom half of the window to form a moisture, insect, and air barrier.

In the other half of the window, there are the lower stiles, which are counterparts to those in the top half of the assembly. The bottom rail is installed between the stiles, just as the top rail was in the upper half. At the top of the bottom half of the window is the lower meeting rail, which is lapped by the upper meeting rail, described above. (See Fig. 9-1.)

The entire sash, or window assembly, can be constructed on-site or it can be ordered. In either event, it should be $\frac{1}{8}$ inch narrower than the window frame and $\frac{1}{16}$ inch shorter. Some modern builders prefer a clearance of only $\frac{1}{32}$ inch in the length of the window. This is literally cutting it pretty close, however, unless you have a perfectly framed window.

If you are buying your window sash, you will probably receive the $\frac{1}{32}$-inch clearance. If it doesn't fit, minor adjustments can be made easily.

To install window sash units years ago, sash weights were always part of the assembly. Modern window tracks, however, have virtually eliminated the use of and need for sash weights.

A *double-hung sash* is simply two window sections installed in one frame, so that each can be raised and lowered. The window unit in the typical house in this country is a double-hung sash.

In such windows each sash must have its own runway. The upper sash slides in the outer runway, while the lower sash moves in the inside runway. In older styles of constructing and installing window sashes, there were strips of wood called *parting stops* which were used to keep the window sashes separated. Pulley pockets were also present on both sides of windows and inside the facing. Modern window construction has eliminated these elements, for the most part. Tension now does the work of the weights and pulleys. Similarly, sash cords have disappeared. Their loss also eliminated a great deal of work and trouble for the home owner and the carpenter. (See Fig. 9-2.)

INSTALLING MODERN WINDOWS

When you install modern windows, start by lifting the top sash—or top half of the window assembly—and holding it inside the frame to see whether you have a good fit. In this case, a good fit is a slight clearance; you do not want a snug fit. If the sash is obviously too tight, you can plane down the sides of the stiles until you have reduced the sash to its proper size. Be careful, however, that you do not plane off too much or that you do not gap or pit the stile by gouging with the plane. Run the plane gently and smoothly along the length of the stiles until you have reduced them uniformly.

Do the same with the bottom sash. Adjust until you have a snug and smooth fit and easy movement.

If the parting stop has already been installed onto the window jambs, remove it and place the upper sash into its proper location and then reinstall the parting stop. Push the upper sash all the way to the top of the window opening and hold it there while you or an assistant drive a small nail under the bottom rail and into the jamb to hold the sash in place.

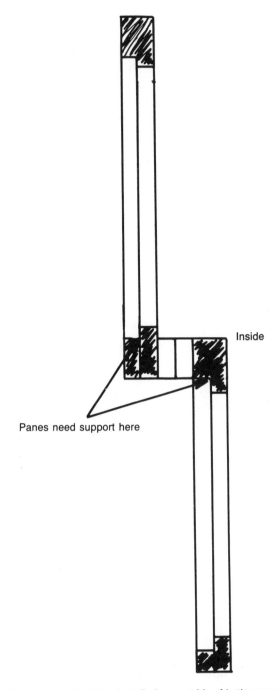

Inside

Panes need support here

Note: Glass panes should be installed on *outside* of both upper & lower sash

Fig. 9-1. When you install windows, make certain that the upper meeting rail and lower meeting rail line up perfectly. If they do not, the window will rattle and admit a great deal of cold air and moisture.

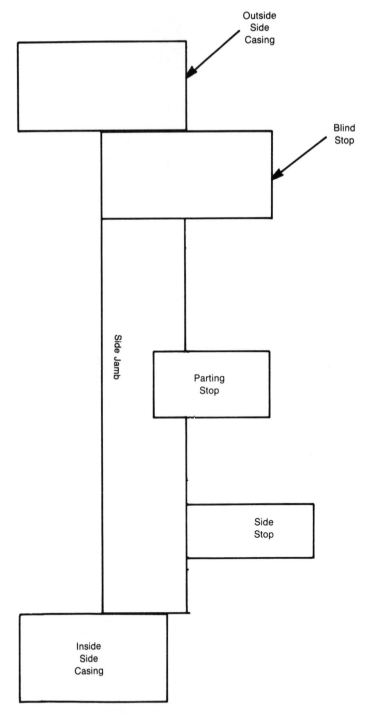

Fig. 9-2. This illustration shows the basic parts of a window from a side view. Each side is composed of the same components in the identical location.

Set the bottom sash in place and push it all the way to the bottom of the frame. Check the meeting rails halfway up the window opening. This is the point where the two sashes overlap slightly. The meeting rails should be flush.

If they are not, but the difference is slight, you can correct it by planing the bottom rail. If the difference is great, you may have to readjust the window frame.

You can buy, at nearly all supply houses, metal tracks for wooden windows. This is the best purchase you can make when it comes to installing windows. These very inexpensive tracks eliminate not only the sash weights and pulleys but also the spring balances, parting stops, notches, and all the other technicalities that made window installation such a difficult job years ago.

To buy the tracks, you must first measure your window assembly carefully. Take the dimensions to your dealer, who will direct you to the proper track for your needs.

These tracks come in two pieces, which can be installed with great ease. One of the easiest ways to install the tracks is to nail one track unit into place by using very small nails or tacks with small heads. Be sure the heads are flush with the metal surface so the window sash will not catch on it when the sash slides up and down.

You will notice that the two tracks in each section are divided by a half-inch parting stop which serves to keep the window sashes separated. When the first half of the pair of tracks is in place, you can slip the stile of the upper sash into the outside track and hold it in place.

While holding it, insert the stile of the lower sash into its designated track also. Now have a helper slide the other half of the track onto the edges of the stiles while you pull the sashes slightly forward so that the edges of the stiles clear the jamb of the window by 3 or 4 inches.

Now gently ease the sashes, with the other track in place, into the window frame. Slide the bottom sash up carefully and then nail the bottom of the track to the jamb. Slide the bottom sash back down and then slide the top sash down also, so that you can nail the top part of the rail to the jamb. That's all there is to it. The sashes are installed.

If you prefer, there is an even easier method. You can buy the sash assembly, complete with tracks already assembled, and you need to do nothing but nail the tracks to the jambs.

MAKING YOUR OWN WINDOW SASH

If you feel that you would like to make your own windows, you can construct a very presentable assembly with only a few tools and some expertise in using them. Admittedly, you will not save any time and very little money, because the time needed to construct the windows could be used in other necessary tasks in building the wall or entire house. You can buy the entire window assembly that requires only your setting it in the window opening and nailing it in place, then adding the trimmers and facing.

If your own satisfaction requires doing it yourself, you will need some wood that is 1½ inches thick and 2 inches wide. For stiles, cut two lengths of the lumber 28 inches long. Be sure to square the edges perfectly. Cut two more lengths 31 inches long. (You can vary the dimensions according to your own needs, but the dimensions given here will provide a large window suitable for bedrooms, dens, or dining rooms.) The 31-inch lengths will be used for bottom rail and lower meeting rail.

Join the units of stock by standing one length of bottom rail on its end. Place one of the stile lengths so that the end of the stile and the end of the bottom rail are perfectly flush. Now drive one finishing nail through the stile and into the bottom rail. Lay the joined pieces flat and use your square to determine that the joint is perfect. When you are satisfied, drive one more finishing nail into the joint. Be sure that the heads of both of the nails are flush with the surface. (See Fig. 9-3.)

Lay the assembly flat and join the other three corners in the same fashion, stopping after each joint to square the assembly, if necessary. When you have finished, you will need to support the frame thoroughly. One way to do so is to buy four small corner braces and screws to fit. Install the braces, one in each corner, so that the windowpane will actually touch the inside edge of the brace.

Another method is to buy *corner brads*, which may be referred to by several other names. These are small lengths of corrugated metal with one edge sharpened and the other flat. To use these brads, stand one with the sharp edge crossing the joint line of the stile and bottom rail. The frame should be laid flat and the corner brad should be driven into the wood so that half of the brad is on one side of the joint line and the other half is on the other side.

When you have driven the brads into each corner, for more support you can stand the frame on its edge and add a brad on the top edge at each joint line. Then turn the frame over and do the same thing on the other end.

Another good way to brace corners is to lay the frame flat and drive one brad in across the joint line diagonally so that one end of the brad is ½ inch from the outside edge of the bottom rail and the other end is 1 inch from the end of the stile.

Add a second brad so that one end is ½ inch from the inside edge of the bottom rail and the other is ½ inch from the end of the first brad. The two brads will form an open V in the stile portion of the joint.

Do this on all four joints, and you will have a very sturdy outside frame for the window sash. This window sash will hold nine 8-×-10 panes of glass, and the upper sash will be the same size.

When you have driven the brads in flush and the window sash frame is painted, the brads are barely visible. If you are concerned about the possibility that the brads can be seen in spite of curtains or drapes, you can turn the sash so that the brads are to the outside.

Your next step is to cut two lengths of trimmer 24 inches long, ¼ inch thick, and with a width the same as the thickness of the sash frame, which is sold as 2-inch wood but in reality has a smaller finished thickness. If you want greater strength, you can cut the piece ½ inch thick and make the panes ¼ inch narrower.

Measure from the inside edge of the stile to a point exactly 10 inches away and mark the edge of the bottom rail. Mark the lower meeting rail similarly and then use finishing nails to nail in the first partition, using the wood you have just cut.

Measure over exactly 10 inches from the first partition piece and mark the bottom rail. Then do the same for the lower meeting rail, and then nail in the second partition piece.

When this is done, cut three 10-inch sections from the same stock you used for the first partitions. Measure exactly 12 inches from the bottom rail. Mark the first partition and then measure and mark the stile correspondingly. Do the same on both sides of the sash frame, and then install the divider pieces you just cut by using either a good strong wood glue or very small nails with small heads.

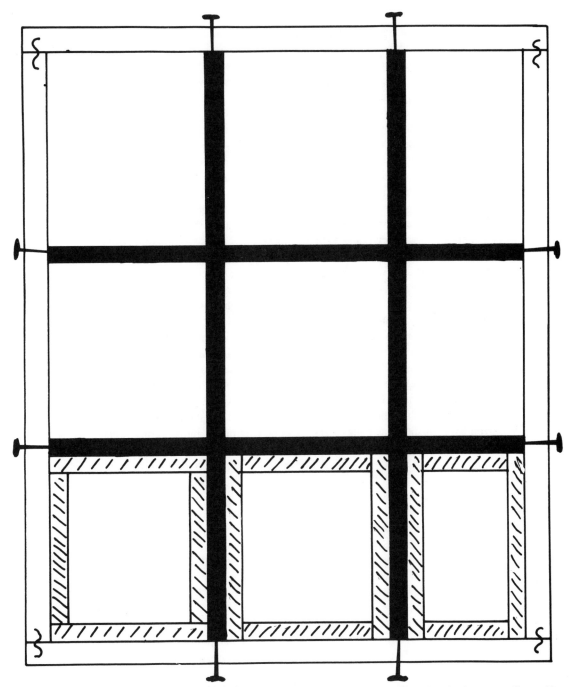

Fig. 9-3. After you have sawed the two pieces, one for the stile and the other for the bottom rail, position them so that the bottom rail butts into the stile. Fasten them by driving a nail from the side of the stile into the end of the bottom rail. You can use screws, angle braces, or corrugated brads, if you wish.

You now need nine inside trimmer pieces that are 10 inches long, ¼ inch thick, and ½ inch wide. Tack or glue these in place between the partitions on top of the bottom rail and on the top and bottom of the middle partition.

Next, measure and cut the next trimmers exactly the same width and thickness as before. The length should be precisely that of the space between the bottom rail and the middle partition and also the same as the distance between the middle partition and the lower meeting rail.

These can be installed with either glue or very small nails vertically against the stiles on either side and on both sides of the partition pieces.

This may sound like a very long and involved activity. Even though the first frame may take a fairly long time to build, however, subsequent frames move much faster.

When this part of the frame is completed, you are ready to turn the assembly over and lay the windowpanes in place. They should fit inside the rectangles allotted for them and they should have a slight amount of clearance on both sides and on the top and bottom.

Install one pane of glass. Then, using a staple gun or small tacks with medium-sized heads, set the pane temporarily, until you are ready to glaze.

To glaze the window panes, first apply a bed of putty along the inside of the trimmer pieces you have installed. The putty bed should be no wider than ¼ inch. Lay the pane so that the inside edge of the glass rests upon the putty bed. Then, with the pane in place, make a roll of putty or glazing compound 12 inches long and lay it against the pane of glass and the edge of the partition piece. Then use a putty knife to smooth the putty.

Hold the knife at such an angle that one edge is against the pane of glass and the other against the partition piece. Start in the corner and pull the putty knife toward you slowly as you apply enough pressure to spread the putty smoothly and to force it into cohesive contact with the pane of glass and the wood surface. At the end of the bead cut the putty cleanly and squarely with the edge of the putty knife. Do the three remaining seams. When you have finished, the pane of glass will be windproof, moistureproof, and rattleproof.

You can leave the small nails you used at first to hold the pane in place. The putty will set around the shaft and head of the nail and will make a long-lasting bond. You can use *glazier's points* instead of small tacks to hold the panes of glass in place. These are tiny triangles that are installed by using a gun similar to a stapler, and they are very easy to apply.

Before you do any type of glazing, you should always clean the surface thoroughly. You do not want any grease, oil, dust, dirt, or sawdust under the putty or glazing compound.

You can see that constructing a window is time consuming, and you will probably benefit greatly by buying the factory-assembled window sashes and frames. Again, the job sounds, on paper, much more difficult than it is.

If you have a pane of glass that does not fit and you need to reduce the size of it slightly, buy or borrow a *glass cutter*. This little device is very inexpensive, and it works wonders with a pane of glass. To use one, lay the pane of glass to be cut on a firm surface, preferably a table top with a thickness of cardboard or similar material under the glass. Position a straightedge where you want the glass cut, and press firmly down, hard enough to hold the straightedge in place but not hard enough to crack the glass. Set the edge of the cutter on the glass surface. Bearing down enough to score or mark the glass in a solid line, pull the glass cutter along the edge of the ruler or straightedge only once.

It is important that you mark or score the glass only one time. The glass has a tendency to break—and at the wrong places—if you score it more than one time.

After you have scored the glass, hold the large part of the pane—the part you intend to keep—in one hand, and be sure to wear gloves while doing this. In the other hand you can either use a pair of pliers or you can hold the strip to be broken off in a gloved hand. Apply pressure with your fingers along the scored lines, and the glass will snap cleanly and perfectly along the scored line.

Another way, and this may be safer, is to lay the pane of glass on the edge of the cutting surface and let the part to be broken off extend off the surface. The edge of the table or other work surface should match the line of the glass. You can hold the pane solidly by placing the heel of a gloved hand on the pane, and you can use pliers to break off the unwanted portion.

Another method that works well is to take two smooth pieces of wood, preferably 2 inches wide and 12 inches long. Hold the wood so that it sandwiches the part of glass to be broken off. Let the edges of the wood align with the scored mark, and then you can hold the wood with your gloved hand and apply enough pressure to snap the glass. The strips of wood will protect you from any splinters of glass and from jagged edges.

If you have never cut glass, you will be amazed at how easy and how accurate it is. You will find that you are not restricted to cutting only straight lines: you can cut half-circles, curved lines, and, in fact, nearly any shape you want.

The trick is to keep the scored line unbroken. You can hear the cutting wheel biting into the glass with a faint crunching sound. If you do not hear this sound, you are not scoring deep enough. If the scoring line is broken for a length of ½ inch or even less, the glass will not break cleanly along the line. If your glass does not break cleanly and you end with a shard that protrudes ½ inch or more, you might as well discard the pane of glass and cut another one. If you try to score and break off the projection, usually you will succeed only in cracking the glass. If you attempt to pinch off the piece with pliers, you will damage the glass and risk injury.

One handy trick is to use a straightedge and mark off the work surface in inches. If you use an underlayer of cardboard, lay it flush with the front edge and one side edge of the table or work surface. Tape it securely in place. Then mark off a grid in inches in both directions so you can place the glass pane on the surface and see at a glance where you should be cutting the glass. You eliminate guesswork and the time- and error-probability by measuring each pane separately and marking.

Some workmen prefer to use a thin coating of turpentine, oil, or kerosene on the glass surface before beginning cutting. When all panes are cut, return to window construction.

Add trimmers at this point, as you did with door openings, and install the sill. You may want to wait to add the casing, apron, and stool.

If you buy your windows, you can expect to pay at least $60 per window and probably much more. If you want to, you can build a rustic window for less than a dollar. Saw two 2-×-2-inch sections the length you wish the window to be, and two more similar sections the width you want the window to have. Fasten the sections into a square, as shown in Fig. 9-4 and add center sections so that you will have space for four large panes of glass.

Fig. 9-4. If you want to save a great deal of money, you can even chain saw your window parts. Shown here is a window frame with the outside portions completed and the inside portion divided into four parts.

Fig. 9-5. The window frame is complete at this point. You can see that a huge diamond occupies the central part of the window. Inside and outside the diamond there are eight triangles. All that remains is to install the glass panes. You could do so by cutting triangles or by buying four large panes and installing them behind the muntins. Total cost of such a window, excluding panes, is less than 20 cents.

If you wish to elaborate on the simple structure, cut and fit partition sections so that all pane sections are divided in half diagonally. A huge diamond is thereby formed in the center of the window. Now you have a modified chalet look. (See Fig. 9-5.)

In the windows shown here, a chain saw was used to cut all lumber, starting with small trees, and all trim. If you use a circular saw, you can achieve a smoother effect. The windows shown here were made for a rustic structure.

With either homemade or factory-made windows, no matter how carefully you install them, there will perhaps be some rattling or sticking or shifting as the house settles. You can often make the corrections in a short time and with a minimum of work. Chapter 10 offers suggestions for correcting such problems.

10

Correcting Door and Window Problems

NO MATTER HOW STURDILY THEY ARE BUILT AND HOW GREAT THE MA-
terials are, all houses sooner or later will shift slightly. The shifting or settling can be
attributed in some instances to foundations that were not dug below the frost line. When
the weight of the house sits on the foundation walls for months or years, the house will
sink anywhere from a fraction of an inch to an inch or two, sometimes more.

CAUSES OF DOOR AND WINDOW PROBLEMS

Settling can be caused in other ways as well. Footings not poured deep enough can
be one cause. If you built your foundation walls on top of *green footings* (concrete not
allowed time to set properly), settling can occur.

Other causes include rotting sills and joists, green lumber for studding and corner
posts, and a host of other possibilities. The important consideration is that settling of the
foundation, even if it causes only a slight deviation from the original position, can produce serious structural problems. For example, suppose that an 8-foot corner post deviates
only ¼ inch from plumb, 12 inches from the soleplate. At the point where the post joins
the top plate, there can be well over 1 inch of variation. One inch is enough to cause
a door to refuse to close or refuse to latch. It also can cause windows to become jammed
hopelessly.

The settling of a house is not always the cause of malfunctioning windows and doors,
though. At times the difficulty is a very simple matter that can be corrected with a
screwdriver or an oil can. Always look for the simple solution first. If a door squeaks,
try a drop or two of oil on the hinges. If a lock is tight or will not turn, you can try lubricating
it with graphite.

Unfortunately, many problems cannot be solved so simply. One of the worst problems is that of dragging doors, and here you may have to spend time and energy to correct
the problem.

Whereas many dragging door problems are caused by sinking foundations, there are other causes. One cause is that of a pier or foundation that is slightly too high, rather than too low. Here's what can happen. If a pier is built on top of a properly prepared footing—that is, one that is dug out until it is fully through the top soil, and then a layer of sand, then gravel, and then 3 inches of concrete is poured—and if the pier is 1 inch or so too high, it is unlikely that the pier will permit any substantial settling. As the house construction progresses, great weight will be placed on the foundation and on that one pier that is too high. Great weight is also placed upon the sills and joists. Wood is more pliant than concrete. Therefore, when the house is complete, its weight will cause the floor and wall supports to give slightly on all sides of the pier. The result is that the wall may be slightly higher on the hinge jamb of the door than it is on the lock jamb.

What this means is that, since the lock side is lower, when the door swings open the outer edge of the door will strike the elevated floor. The hardwood floor will be severely damaged by the bottom of the door, the door itself can be damaged, and the carpet, if any, can be ruined in that portion of the room.

The problem is compounded if you take the door off its hinges and saw off part of the door so that it will open easily all the way. When the door is closed there is a huge crack under it.

Another problem that occurs regularly is that, as the house or part of the house shifts, the door framing can be pulled slightly apart at the top, bottom, or both locations. The result is that the door latch does not make proper contact with the lock and it will not stay locked or closed.

The reverse of this same problem can occur also. If the frame of the house shifts enough to cause the door frame to shift inward rather than outward, the door will strike the facing and will not close all the way. Or, in a similar problem, it will close, but the fit is so tight that the door can be opened only with difficulty. Particularly in wet weather, it might not open at all.

CORRECTING DOOR PROBLEMS

You can solve simple problems easily. If a door squeaks when it is opened or closed, or if it makes a harsh grating sound, you can correct the problem by putting oil on the top of the hinges and then working the door back and forth rapidly. Usually six or seven rapid movements of the door will allow the oil to penetrate into the binding part of the hinge and lubricate the hinge pin so that it moves fluidly inside the hinge.

If this solution is not sufficient, you can remove the pins, one at a time, and dip them in oil. Wipe them lightly before you reinsert them. If the door noise problem persists, remove each pin, again one at a time, and swab shortening, lard, or grease generously over the pin, then replace it.

If a door lock squeaks and grinds, this can often be solved by buying graphite at your local hardware or supply house and applying it to the lock bolt. While applying it, either turn the knob slowly back and forth, if it is that type of lock, or turn the key slowly. Repeat this process until the squeaking stops. If the noise persists after a minute or so of application, further steps are necessary.

One of those steps is to remove the long screws on either side of the door in the lock assembly, if it is a dead bolt lock, and disassemble the lock carefully. Then apply either graphite or a thin film of oil, before reassembling the lock.

If the lock is one in which the turning of the inside knob releases the lock, you can remove the knob and shaft and then add graphite to the inner workings. If this is difficult, apply a small amount of oil with an oil can.

Occasionally, particularly in newer houses, a door knob will rattle annoyingly. You can try to solve this problem by using a small screwdriver to tighten the tiny set screw that you will find (between the knob and the door). It is this screw that holds the lock in place on the shaft. It is probable that the set screw was not tightened well when the door was hung.

If tightening will not work, remove the set screw and take the knob off the shaft. You will find an opening in the door knob where the knob fits onto the shaft. You can insert a very small amount of putty or modeling clay into the hole. Try to keep the clay or putty just inside the opening and applied to the edges of the shaft hole only. You can apply the putty easily with the point of the blade of a small screwdriver, the tip of the blade of a pen knife, or the point of a letter opener. Another very handy tool to use is the stick from an ice cream bar. This is probably the best device you can use for applying the putty.

Return the knob to the shaft and push it in as far as it will go. Then tighten the set screw firmly, and the doorknob rattle should be eliminated.

If the door will close completely but tends to stick or bind so that it is hard to open or it opens with a sudden jerk, you can handle the problem easily. First of all, check the hinges to see that they are holding the door firmly. Make a visual inspection first. Then use a screwdriver to test the screws. If they are loose, tighten them and then try the door to see if it still binds or sticks.

Sometimes you will find that the screws will not tighten. What has happened here is that, first, the door has been hung several times. The wood has been chewed out by the insertion of the screws repeatedly and there is no pressure applied by the wood fibers against the screw. The door, then, is simply being held in place by the pressure of its weight against the screws. Second, the screws might have been tightened excessively and in the process the wood was reamed out, leaving nothing to bind the screw. A third possibility is that the screws were too small in the first place and the repeated pressure as the door opened and closed had eroded the wood and the binding power of the screws.

Fourth, screws that were too long for their diameter might have been used, and the thin shaft allowed the screws to bend slightly in the direction of the opposite door jamb. The final likelihood is that the original screws might have been removed and smaller ones inserted in their places.

Solving the problem is actually easier than stating it and its causes. First, try larger and slightly longer screws, but be sure the screw heads fit flush against the hinge. Tighten the screws all the way until the screw bites firmly into new wood and holds securely.

When you have finished, test it by opening the door and standing directly in front of it so you can grasp the knobs on both sides of the door. Now lift, applying enough pressure to lift a 20-pound weight. If the door gives with pressure, you know that your problem has not been completely solved. If it feels tight and sound to the pressure, you have completed the job.

If the door is still slightly loose in the hinges, try an age-old solution. Nails and screws derive their holding power from the wood fibers pushing against the shaft of the screw or the nail. In both cases the shaft shoves wood fibers out of the path as it penetrates.

Instantly these fibers respond by trying to regain their position. The pressure they apply to the shaft serves to hold the nail or screw in place.

A reamed-out hole applies little or no pressure, so it is necessary to restore a pressure source. Do this by using the end of a wooden match, the middle and fattest portion of an ordinary toothpick, or wood putty, sometimes called plastic wood. You can break off ½ inch of the match or toothpick and shove the wood into the hole. Tap the end gently with a small hammer if necessary, and then start the screw as you normally would. The fibers of the new wood will give at first, then exert their own pressure against the screw shaft. The result is a holding power that is virtually as great as the original wood.

If you use plastic wood or wood putty, you can pack the putty into the screw holes, allow it to set up and dry completely and then run the screws in. This process works well much of the time, but it is not universally successful, so you might want to try the match end or toothpick first. The problem with the wood putty is that it is very difficult to pack it into the hole as you need to do, and often you have restored the holding power only a fraction of the depth of the hole. The screw then will crack the putty when the door is used frequently and the door sags again in a very short time.

When you lifted on the door knob and found that the door did not give at all, your problem is not with the hinge but with either a swollen door or shifting foundation or door frame. You can determine the extent of the problem by a visual examination of the edge of the door. You will see a section of the door edge that is considerably shinier than the rest of the door. If the door has been stained or painted, the covering will be either very thin or absent in the trouble spot.

Correct the tightness by using coarse sandpaper on the spot in a gentle sweeping motion. Do not apply excessive pressure and do not concentrate on a small portion of the door edge. Instead, start slightly above or below the trouble area and start a gentle motion across the area to be sanded. As you move your hand, apply slightly more pressure as you approach the middle of the area, and then start to lighten the pressure as you near the end of the area you are reducing. (See Fig. 10-1.)

If the door is too loose in the frame, that is, if there is too much clearance, you have a much greater problem. However, there are several ways you can remedy the difficulty.

If the clearance is too great on the lock jamb side, your method of correction depends upon the degree of clearance. If there is only enough clearance that the bolt or latch will barely catch but not hold securely, it might be that the striker plate is set too deeply into the jamb. If this is the case, remove the screws that hold the plate and lift the plate from its jamb location. Then, with the plate still out, run the screws through their openings and slip a thick washer on the back side of each screw. Then replace the plate and drive the screws back into their original positions. (See Fig. 10-2.)

It is possible now, however, that the screws are not long enough to bite deeply into the wood. In this case, simply use longer screws and retain the washers on the back of the plate.

You might need to use two washers on each screw. The smaller the washer in diameter, the better, as long as you have the needed thickness to force the plate out ⅛ inch or so.

If there is simply too much clearance for the washers to help alleviate the problem, you can cut a very thin slat of soft wood, such as white pine or similar lumber, the exact width of the door's thickness. Attach the slat on the edge of the door on the hinge side. The thickness of the slat should be no more than ¼ inch, less if you can manage it.

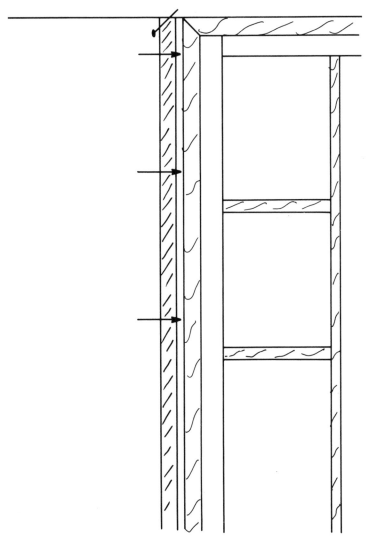

Fig. 10-1. You can generally spot the points where the door fits improperly by making a visual inspection. The door edge will be worn smooth by the tight fit. You can wrap a piece of coarse sandpaper around a block of wood and rub the trouble spots with firm pressure until the door fits as it should.

There is almost always some play in hinges. When you attach the slat to the plate, when you close the door, the added thickness will force the door over so that the latch catches in the strike plate.

You might find it necessary to move the hinges over slightly, if the thickness required is too great. If you do so, you can pack plastic wood or wood putty into the original screw holes and then sand the putty smooth. You can later stain or paint over the putty, which resembles wood to a considerable extent.

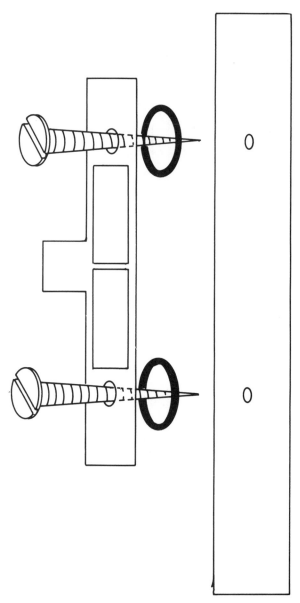

Fig. 10-2. A thin washer on the back of a striker plate will cause the striker plate to be moved forward so that the lock will engage properly. You can use rubber plumber's washers or metal washers.

If moving the hinges is a problem that must be avoided, if possible, one alternative is to removing the casing of the door and add wedge blocks behind the jamb on the lock side. Attach these blocks, which are usually no more than 12 inches in length and the desired thickness, to the wall framing, then replace the casing. You might find it necessary to alter the length of the casing above the door, if only slightly.

The huge problem that was mentioned earlier—that of the door dragging on the floor and damaging hardwood flooring or carpet—is much more difficult to correct. The simplest way, if this is a feasible alternative, is to take the door off the hinges and cut off enough on the bottom to allow the door to miss the floor as the door swings open. Then rehang the door and add a thicker threshold so that when the door is closed the clearance will not be so very obvious.

It is difficult to know just how much to cut off the bottom of the door. You want to remove only that part which is absolutely necessary for proper swinging movement of the door, and you want the cut to be as straight and neat as possible. The best way to deal with the problem is to open the door slowly until it barely touches the floor. Look carefully to see how much of the door now clears the floor and how much is in contact. Mark the point on the door where the clearance starts and, when the door is off the hinges, use a straightedge to mark the cut line and cut off the bottom of the door accordingly.

Now try the door again. If it drags again, mark the spot, remove the door again, and recut.

Some doors cannot be cut realistically, either for fear of permanently damaging a beautiful and expensive door or causing an unsightly appearance. If this is your problem, it is possible to secure a hydraulic jack and go under the house to the pier that is causing the trouble. Build a solid temporary foundation and place the jack on it so that it is securely in place. Then jack slowly until the foundation at that point barely clears the pier. Next, build a temporary pier adjacent to the incorrect one and then use a chisel and hammer to remove the top course of bricks on the first pier.

When this is done, replace the course with a thinner one and then remove the temporary pier. Release the pressure on the jack slowly and gently until the foundation rests again upon the original foundation pier.

Warning! This is a very delicate operation and should not be attempted except with adequate equipment and expertise. If you are not certain you can handle this work competently and safely, hire someone to do it for you or solve the problem in one of the ways mentioned above.

The most important item in such a task is obviously the jack. You must use one that is able to handle the load that will be placed upon it. When the jack starts to lift the immediate portion of the foundation, be certain that the jack remains perfectly vertical. If the jack starts to lean, even slightly, release the pressure, correct the problem, and start again.

Under no circumstance should you permit any part of your body to be between the foundation and the pier. Again, if you are not fully competent, hire someone.

Sometimes a door, when closed, has more hinge clearance at the top than it has at the bottom, or vice versa. In such a case, the easiest solution is to make a visual inspection of the problem and determine whether the hinge is set incorrectly on the edge of the door or on the jamb. If the problem is on the door, you will need to remove the door from the hinges. Mark a slightly deeper hinge cut, with the added space in the direction of the lock side of the door, and reinstall the hinge.

Sometimes a door will not clear the closing area in only one part of the opening. When you try to close the door, note whether the problem is at the top or bottom. This is rare, but there are times in which the door actually clears at both top and bottom but not in the middle.

When you have determined where the problem is, check the jamb for plumb. If it is out of plumb, remove the door casing and make the correction by replacing the wedge block with a thinner one. Before replacing the casing, try the door several times to be sure that the clearance is now correct. Then replace the casing.

Sometimes a door is hung perfectly—apparently. Yet the door will not close properly. When you make the necessary checks you find that the jambs are all plumb, the hinges are installed in every way correctly, and the framing is square in all corners. But the problem persists.

The difficulty here is, quite likely, that the door closed correctly when it was hung, but moisture in some way penetrated the wood of the door and caused a slight swelling. In this case, light sandpapering is the best solution. When you have completed the sandpapering and you are confident that the door is now dried fully, use a sealer and then paint the door to prevent further moisture damage.

CORRECTING WINDOW PROBLEMS

Windows often cause about as much trouble as doors. Windows that worked perfectly when they were installed later refuse to work at all. Sometimes they refuse to rise, even under great pressure, or they go up easily and then will not come down.

There are a few fairly simple ways to attempt to correct the difficulties. One way is to apply a thin coating of oil in the window tracks, if you have used metal tracks. As soon as you apply the oil, force the window up slightly, then lower it. Repeat the process, but this time try to get the window slightly higher than the previous time. As you get the window higher, apply the thin coating of oil both above and below the window sash. Do this on both sides of the sash—in both tracks.

If the windows have the older wood tracks, you can free stuck windows often simply by hitting the window sash frame with the heel of your hand or by placing a length of 2 × 4, end-first, against the window and hitting the opposite end with a hammer. Hit sharply but not hard enough to break or crack window panes. To keep windows sliding freely, sandpaper the tracks until they are perfectly smooth before you apply paint, varnish, or stain. If a piece of wood is not smooth, when it is painted or stained the tiny loose fibers will collect droplets of paint. This will result in a paint build-up in the area. When you have only a slight clearance to begin with, a slight build-up on both sides of the sash will soon eliminate the clearance and the result is stuck windows.

One method that has been used for years to free windows that went up but refused to come down is that of laying a short 2-×-4 piece on the top of the sash—the top rail—and hitting the 2 × 4 with a hammer until the window is freed.

The problem with this dubious solution is that often the pressure is applied to the part that is already stuck. The extra pressure causes the window to be forced into a crooked position inside the frame. The problem is made worse, therefore, by the attempted solution. Worse, hitting the 2 × 4 too hard and too often can cause the window rails and stiles to separate, the putty or glazing to fall out, and the panes to crack.

If you are going to try this solution, find a length of 2 × 4 that will reach nearly all the way across the top rail. With the lumber in position, tap gently along the entire length of the 2 × 4. In this fashion you urge the window gently down, and you do not risk ruining or badly damaging the window itself.

Before you try any of the methods, again we urge you to use sandpaper to clean away all of the collected paint, flakes of loose paint, and fiber ends that could cause the window to jam.

Ultimately, you will probably need to take down the window casing and remove the window. To do this, use a small crowbar or pry bar, a thin block of wood, and a hammer with claws. Insert the point of the crowbar under the window casing and, before you pry up at all, lift the curve of the crowbar slightly so that you can put the thin block of wood under the curve. For best results, the block should be at least 6 inches long and 4 inches wide. The purpose of the block is to keep the pressure from the curve of the crowbar from damaging the wood or paneling on the side of the window.

When you are ready, apply slight but steady pressure on the crow bar handle until the nails give slightly. Now insert the claws of a hammer in the space and move the crowbar several inches away. Using the block again, urge the nails further out. Use this pattern along the length of the side casing until the piece can be removed. Do the same with the remainder of the casing, and then lift the entire window assembly from the opening.

With the window frame out, free the window sash and then make the necessary adjustments. Either sand the entire side of both of the stiles so that you are confident the clearance is enough, or make the necessary adjustments in the framing of the window.

Before you replace the casing, after you have reinstalled the window assembly, drive all the nails back through until the points are flush with the inside of the casing. Then try to locate the same nail holes that were used before and reattach the casings.

If your windows rattle during high winds, even though you have used metal rails, you can usually correct the problem by checking the tracks to see if the retaining nails have worked loose in the wood, thus allowing the tracks to move easily under slight pressure. When you have found the source of the trouble, renail, using either slightly larger nails or add more to the loose track end.

Windows with wood tracks tend to rattle very easily, and the solution here is to find which side of the sash is loose. You can make this determination by pushing on both stiles, and if the window gives excessively, you have found the problem. Or you can raise the window 6 or 8 inches and grasp the bottom rail and push and pull. Try both sides to determine where the excessive clearance is.

Older windows have wooden strips called parting stops that run vertically along the middle portion of the side jambs and keep the upper sash from bumping into the bottom sash. It is likely that the parting stop on one side or the other had been installed with too much clearance, and you will need to remove the parting stop and move it over, at least on one side, until the clearance is corrected.

Because the nails used to hold parting stops are very small, you can remove the small slat with the blade of a medium-sized screwdriver without damaging the wood. When you have moved one of the parting stops, try the window again to see if the rattling has stopped. If so, your job is nearly completed. If not, you might need to move the other parting stop.

In any event, you will have solved one problem and allowed another to start, for if the parting stop was moved toward the inside, you left too much clearance for the top sash, and it will now be free to rattle. If you moved the parting stop toward the outside, you have made the top sash too tight to move freely and easily, so one final step remains.

Between the outside casing and the top of the upper sash is a strip of wood called the blind stop, which fits flush against the side jambs. You may need to free this blind stop and move it either in or out slightly, depending upon where the clearance problem is located.

Sometimes a window sash will actually tilt from side to side slightly inside the tracks. Such a problem can cause the window to refuse to stay up when raised, or it can allow the sash to become stuck. To free a window that is stuck in this manner, check first visually to see which side of the sash is lower than the other. If you can't tell visually, place a level on the top rail to help you to determine the low point. If the sash is raised far enough, use the head of a hammer and gently tap upward until the bind can be eliminated. Then the window sash should slide up and down freely, but it still may be too loose.

Next, place a level along the side jambs to determine which of the jambs is not plumb. You will then need to remove the casing lumber and window assembly so the jamb problem can be corrected. You can make the correction in the same way you corrected the jamb plumb problem in the door frame.

You will need to remove the jamb and nail a small wedge block against the studding so that the jamb plumb line is corrected. Be careful not to use a wedge block so thick that it will make the window track so tight that the sash cannot move freely.

The following chapter will show you, in a step-by-step demonstration, how to construct a wall frame for a wall that serves both as a partition and as a load-bearing wall.

11

Wall Framing
Unassisted:
A Sample Problem

IN THIS CHAPTER YOU WILL SEE HOW ONE PERSON, WORKING TOTALLY unassisted, can handle the problems of wall framing. You will follow each step in a logical sequence until the walls of a pair of rooms have been framed.

EVALUATING A BUILDING PROJECT

In this case, the roof is already on the structure, and 4-×-4 posts hold up the roof. But the huge roofed-in area now is to be converted to a pair of spare rooms and a small sitting porch.

The purpose of this chapter is to show what can be done to solve several problems. In the instance shown, the area was originally a deck. The family had little use for it, however, and the flat roof that covered the sitting area of the deck leaked. After trying repeatedly to locate and stop the leak, the family decided to roof over the deck and then, later, to modify the space into usable rooms.

The problem was therefore two-fold. The roof was already in place, which meant that it was virtually impossible to install top plates and caps the way you would ordinarily construct a wall frame. The bearing walls would have to be inserted under the roof, so the partition wall could also help support the long roof. The partition wall between the two rooms would also help support the roof line. This chapter demonstrates how these problems can be solved easily and advantageously.

The deck that was to be rebuilt occupied a large expanse of the top story of a two-story Early American house. The house, in fact, was built prior to the Civil War and the area of the deck was extremely unusual for the times. The modern owners of the house decided that it was not practical for more than 600 square feet to be devoted to a sitting area, which was used only occasionally. They therefore decided to modify the space.

Their decision was also influenced by the fact that a leak had allowed rain and melting snow to start to seep into the rooms below. After several futile attempts to patch the leaks, the owners decided to roof the area, more to stop water damage than to add space to the existing square footage of the house.

They decided that the area could serve as an office for the family business or double as an efficiency apartment. Floor plans, therefore, were hastily sketched on a sheet of notebook paper.

The ends of the roof supports on one side of the house were supported by the peak of the roof that covered the kitchen, dining room, and servants' quarters areas of the house. The other ends of the rafters had to be supported by the 4-×-4 posts that were erected across the front of the deck. The new plans called for a portion of the deck to be set aside as a screened-in sitting area for the upstairs portion of the house.

If you have a similar problem in planning the best use of existing space, this example may be instructive. Many people plan alterations to their home in light of their current, rather than future, need. This chapter demonstrates how building plans can be modified to accommodate these future needs.

If you have a sitting porch, therefore, as opposed to a stoop, that is not really practical, or if you have a deck, covered patio, or any other type of space that is needed more for a new room, you can modify the approach used here to fit your own needs.

In this case one person did the entire job. You will see how you, working alone, can do the same thing. It is not always easy to work alone, but at the same time it is not always possible to pay for assistants.

BUILDING THE BASIC WALL FRAME

Start building your load-bearing walls, as always, by marking off the line you want the wall to follow. Use a chalkline so that the line will be clear, straight, and perfectly in line with the rest of the space—in this case, the deck space. If you are dealing with a defined limit, such as the edge of the foundation, the location of the bearing wall is no problem: it is dictated to you by the conditions.

In this instance, measure in from the edge of the deck to the point where the sitting porch would stop. In this case it was agreed that a 54-inch-wide porch would be adequate, since heated space was more important than fresh-air space. Measure in 54 inches, therefore, from the edge of the deck at both ends. Then chalk a line connecting the two points. (See Fig. 11-1.)

Normally you would mark the line and then proceed to build the wall frame, but one person would have difficulty in lifting a wall frame as long as this one—nearly 30 feet—so the first step is to select and cut the soleplate for the wall. If you use one long 2 × 4, take care that it does not bend or sway slightly in the middle. No matter how straight a timber is when it is cut, the position in which it is dried determines whether the timber retains its trueness.

If the timber is crooked, start at one end and align the timber for the soleplate with the chalked line as far as you can. Nail the first end in place by using two 20d nails.

Then move out 2 feet and check the soleplate for position. If the outside edge does not align perfectly with the chalked line, you will need to tap it into position by hitting it gently on the edge with a hammer. Pay no attention at this point to the location of the other end of the soleplate. It may be as much as 2, 3, or even 4 feet out of alignment.

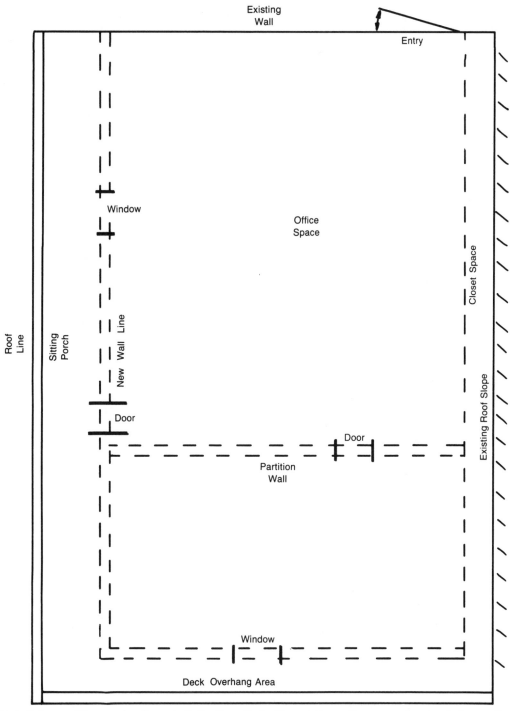

Fig. 11-1. Before cutting any lumber, chalk the lines that will mark the wall you are planning to frame. Examine the wall line for any future problems affecting plumbing or wiring.

You are concerned only with the alignment of the first two feet at this point, however. So when the timber is positioned correctly, fasten the soleplate at that point with two 20d nails. Then move out 2 more feet and align the edge of the soleplate with the chalkline at the 4-foot location.

Follow this procedure the entire length of the soleplate, gradually forcing the timber into conformity with the chalked line as you go. The soleplate is now installed, and you are ready to move to the corner posts.

If you are using shorter lengths of 2 × 4s for the soleplate, you should not have any trouble with straightness. You can pick your best timbers, therefore, and nail them into position.

Remember the function of the soleplate. It is the timber upon which the entire room will rest ultimately. If it is weak, damaged, or irregularly milled, you are inviting troubles in the studding and capping.

Before you nail down the soleplate made from the shorter timbers, lay off exactly where the studs will go. Position the soleplate lumber where you think it will go, and then start at the end and mark off 4 inches. This space is for your corner plate.

Measuring from the end of the timber, not from the end of the allocation for the corner post, mark the location for the first stud. Mark all stud locations for 16-inch on-center installment, so you can add sheathing or siding and insulation later with the greatest of convenience.

When you come to the final stud location for the first soleplate timber, be sure that you will not have a stud located on the very edge of the soleplate joint. Remember that a stud located on a joint has less than 1 inch of support from each section of the soleplate, since the stud would actually straddle the joint. And if the end of the soleplate section is weak, then you will have a defective studding situation from the beginning. If you need to do so, cut the soleplate timber so that the joint will come at the mid-point between studs.

Later, if you wish, you can cut short lengths of 2-×-4 lumber to nail over the joints for added strength. Lapping the joints will not interfere at all with later insulation or wall-covering work.

You probably have already noted a future difficulty with the building of a bearing wall under an existing roof. If you erect the corner posts now, how can the top plate be installed? And if the roof is not perfectly constructed, how will you know that the top plate, even if you can figure out a way to nail it in place, will be right for the corner posts?

Here is the solution. Measure the precise distance from the soleplate to the rafters, then deduct from that distance the thickness of two 2 × 4s—one for the top plate and one for the cap. There is no practical way you can nail up the cap before the top plate is in position, and you cannot nail in the top plate before the corner posts are erected.

It is rather difficult for one person to set up the corner posts, but it can be done. After you have constructed the first corner post, put it aside until you can cut two brace pieces, each about four feet long. The length is not crucial, so if you have some sections of 1-×-4 lumber handy, use them, as long as they are 3 feet long or longer.

Lift the corner post and position it as it will be installed. Nail one brace piece to the left outside (assuming that the corner post is for the left front corner, as you face the wall), and the other brace to the front. The first brace extends along the left wall line, and the second brace extends along the front right wall line. Nail the braces about 5 feet from the bottom of the corner post. Use only one nail in each brace, and do not drive the nail in until it is flush with the wood. Instead, leave at least 1 inch protruding so you can pull the nail out easily later.

Stand the corner post on top of the soleplate and in the position it will be in permanently. Then extend one brace as far as it will reach while, still holding the corner post in its position, you start to position the second brace.

When the corner post is in what appears to the eye to be a vertical position (actually, it needs to lean inward slightly so it will not be likely to fall), leave it standing while you nail to the floor a short piece of 2 × 4 beside the end of each brace. Take your level with you as you return to the corner post, and when you have the post in a true vertical position, hold the post steady by pushing the brace piece toward the floor. Then follow the brace to the end and nail it to the 2-×-4 length you nailed to the floor earlier.

You now have the corner post true in one direction. You now can drive two 20d nails at an angle from the outside left through the corner post and into the soleplate.

Now use the level on the other side of the corner post to get a plumb reading, and do the same thing you just did. Hold the post in position by pushing down on the brace, and then nail the brace to the floor piece.

At this time you can complete nailing the corner post to the soleplate. Use 20d nails on all four sides, two nails to the side, and leave the braces in position.

Erect the other corner post in exactly the same manner, and leave it braced. Take care that the braces will not interfere with your work room as you complete other parts of the wall framing.

STUDDING OUT THE WALL

With the corner posts nailed in and plumbed, start marking the locations for the studs, partition studs, and rough door opening, if any. Mark the stud locations so that all studs will be exactly 16 inches on center. This distance is crucial, remember, since every three studs equal 48 inches, or 4 feet, which is the width of a sheet of paneling, gypsum board, sheathing of all types, and fiber board. Many rolls of insulation come in 15-inch widths, which is perfect for the space between the studs.

When you are marking off partition studs, do not let the fact that the stud is wider than the normal stud interfere with your 16-inch centers. You are concerned only with the center of the stud, not the outside positions. Now you can nail in the common studs.

Again, it is somewhat difficult to do so, but with a little help from a length of 2 × 4 you can do it quite easily. Measure the exact distance between the inside edge of the corner post and the near edge of the first stud marking. Then cut a piece of 2 × 4 that will fit exactly between the corner post and the near mark of the first stud. Lay the short timber on the soleplate.

Get your first stud and stand it in its marked position so that one side is firmly against the short 2 × 4 you placed atop the soleplate. Balance the stud against your body and drive two 16d nails diagonally through the far edge of the stud and into the soleplate.

Now tap the short 2 × 4 (the stud spacer) out and drive two more 16d nails diagonally into the near edge of the stud and into the soleplate. The stud will not be firm at this point but it will stand erect while you work on subsequent studs.

The stud spacer will not work for the next stud, since the stud spacer was cut to fit between the corner post, which is 2 inches wider than a stud. So measure the distance between the far edge of the first stud to the near edge of the next stud marker and cut a 2-×-4 length that will fit exactly in that space, just as you did before. This second piece will work for all common studs for the rest of your framing work. Use the second piece of 2 × 4 as you did the first, and nail up all common studs on the wall line.

When you have nailed in all common studs, it is time to nail in the partition studs, if any. To make a partition stud, lay one common stud flat and place two other common studs on edge against the edges of the flat stud. Nail the three studs together to form what looks like a rough trough.

When you are ready, nail in the partition stud just as you would install a common stud, but keep the open side of it facing into the room. Later, you will join walls by pushing the first stud in the partition into the open face of the partition stud and nailing it securely. This gives great strength and stability to a wall and makes it well worth the effort needed to build the special partition stud. (See Fig. 11-2.)

You will deviate from your pattern only when you reach the rough door opening. After that point, plan 16-inch centers for all studs, normal and otherwise.

INSTALLING THE TOP PLATE

When the wall is studded, all studs have been nailed to the soleplate only. It is now time to install the top plate. This step can be one of the most difficult you will make.

When you start to nail in the top plates, you face an instant problem: how are you going to nail the plates in place if you cannot get a hammer to the upper side of the plates, because the roof is in your way? The answer is: if you can't nail down, nail up.

When you installed corner posts and studs, you had to cut them to the necessary length to reach the roof rafters, minus the thickness of two 2 × 4s. One of the 2 × 4s you left room for is the top plate.

Your first step in installing the top plate is to locate, examine, and choose lumber that is suitable. If you have one very long timber that will reach from one corner post to the other, you can use it. The only difficulty is that you will have to handle a very long and very heavy timber, one that, while it would not give you great difficulty under normal conditions, will be somewhat difficult to handle while it is perched upon the ends of unstable 2 × 4s.

You can make the job easier by cutting two 2-×-4 lengths, each about 12 inches long. Scrap lumber of nearly any sort will work. Take the scrap lumber to the common stud at the far end of the wall line. Nail a length of the scrap wood to the top of the common stud: one piece on the inside and the other on the outside. Each piece should extend about 4 inches beyond the top of the stud.

Use only two nails on each side, and do not drive the nails all the way into the wood. Leave 1 inch sticking out so the nails will be easy to extract when you no longer need the brace pieces, since the only function of these lengths of wood is to hold the top plate in place temporarily.

When you are ready, lift one end of the top plate and push the end between the braces you nailed to the common stud. Let the top plate extend to within 3½ inches of the end of the wall.

Now go to the other end of the top plate and lift it. Then slip it on top of the other corner post. You may need to pull it toward you slightly until the end is within the width of a 2 × 4 of the outer edge of the corner post.

Go back to the first corner post and check the length of the top plate. If it is within the width of a 2 × 4 from the edge of the corner post, you are ready to nail it in place. If the length is not right, mark the cut line, remove the top plate, and cut it to the proper length.

Drive nails at an angle up through the corner post into the top plate. Put at least two 20d nails in the end, and angle them enough that the nails must pass through adequate wood in order for the plate to be held in place securely.

You can start the first nail, and when it is sunk deep enough, use one hand to hold the top plate in place while you sink the nail flush with the surface of the corner post. The top plate will be steady enough that you can drive the second nail without difficulty.

Fig. 11-2. Nail in the partition stud so that the open side of it faces the interior of the room. The partition wall will eventually be fitted to the partition post.

You might have trouble holding the top plate while the first nail is driven. If so, you can handle this job by nailing a short piece of 2 × 4 across the inside edge of the corner post and flush with the top edge.

With the 2 × 4 in place, use your C-clamp by opening it wide enough that you can put one end of it atop the top plate and the other under the piece of 2 × 4. Tighten the clamp until the top plate is held securely. Then nail confidently while the C-clamp holds the top plate for you.

The other end of the top plate is being held in place by the two braces you nailed to the other corner post. Nail the other end of the top plate in place now, using the same method.

At this time you can nail the tops of the common studs and the partition studs in place. Earlier you cut a length of 2 × 4, a stud spacer, that reached from the inside edge of the corner post to the near edge of the first stud. Use that same spacer now to help you nail in the first stud.

This time, to make your work easier, remove the small scrap of 2 × 4 you nailed across the corner post and hold the stud spacer against the underside of the top plate. Now use the C-clamp again and clamp the stud spacer to the top plate. Then, at the far edge of the stud, push its top into position. Hold it tightly against the end of the stud spacer while you nail the stud to the top plate. Start the nails at an angle and drive them upward through the corner of the stud into the top plate.

Cut another stud spacer, too: one that measures the exact distance between common studs rather than between the corner post and the first stud. Now use the second spacer in the same way and nail the top ends of all common studs. When you come to partition studs, if any, you will still be able to use the C-clamp, but now any length of 2 × 4 will work, as long as you have determined that the stud is plumb.

It is a good idea to check the studs as you nail them in to see that they do not deviate from the vertical. Do the same with the corner posts and make any needed corrections while they can be made with little trouble. You can now remove the braces at the top of the corner post, but do not take down the braces that hold the corner posts in a vertical and plumb position.

BUILDING THE DOOR OPENING

Next lay out the rough door opening, if you have not done so already. The rough opening of a door is usually 2½ inches wider and 2½ inches taller than the door itself. The extra space will be taken up by the jambs and wedge blocks. Assume, therefore, that the door is 32 inches wide and 80 inches high. The rough opening for the door would be 34½ inches wide and 82½ inches high. The inside edges of your door frame studs should be on the 34½-inch mark. (These dimensions are based on the assumption that the trimmer studs and header are already in place and that you still have 34½ inches between trimmer studs.)

It is now time to construct the header for the door frame. If the rough door opening is only 3 feet wide, you can use 2-×-4 lumber to make the header, but if you have 2-×-6 lumber available, use it instead.

Remember to make the header the same exact length as the space between the studs on each side of the rough door opening. You can make the header by standing one 2 ×

4, then laying another 2 × 4 so that the outside edges of the two are flush. Nail the two pieces together and then place a third length of 2 × 4 exactly as you did the second piece. Nail the third piece in place.

You can use another method if you prefer. If your rough door opening is 34½ inches wide, cut two 2 × 4s this exact length. Also cut three sections of ½-inch plywood 3½ inches by 3 inches. Lay one 2 × 4 flat and place one plywood block so that it is flush with the end of the 2 × 4. Do the same at the other end. Place the third plywood block in the exact center of the 2 × 4.

Now lay the second 2 × 4 so it is flush with the first, or bottom, 2 × 4. Do not disturb the location of the plywood blocks that are now sandwiched between the 2 × 4s. Drive three 16d nails into each end and in the center of the assembled header. Be sure that the nails penetrate the plywood wedges and extend through the plywood into the bottom 2 × 4.

Cut trimmer studs at this time. These are normal studs that are regular length minus the height of the header. In other words, if the header is 3½ inches tall, then that much should be cut off the ends of the trimmer studs. If the header is 5½ inches tall, then cut that amount off the normal studs.

Nail the trimmer studs onto the studding in place, with one nailed to each of the common studs marking the boundary of the rough door opening. The bottom end of the trimmer stud should rest on the soleplate. At this point put the header into place by setting it on top of the ends of the trimmer studs. (See Fig. 11-3.)

Nail the header in by driving 20d nails through the studding and into the ends of the 2 × 4s of the header. Drive nails into both of the 2 × 4s on each end, 3 nails per 2 × 4 or a total of 12 nails for the entire header.

Rough in the window opening. Then construct another header, this one as long as the window opening is wide. The trimmer studs for the rough window opening are in two pieces, one below the rough sill and one above it, on each side. Cut the lower trimmer studs according to the height of the rough window opening and nail them to the studs on each side of the rough opening.

Cut the rough sill from a 2 × 4 and make it the same length as the header. Nail in the rough sill on top of the lower trimmer studs.

Now cut the upper trimmer studs, which are from 2-×-4 stock also. They should be the length of the distance from the top of the rough sill to the bottom of the header. Nail these in place as you did the lower trimmer studs.

Now set the header on top of the ends of the upper trimmer studs. Nail them in place by driving 16d nails through the common studs into the ends of the header 2 × 4s, just as you did for the door header.

If a partition is to intersect the wall line, mark the place where the two walls will meet and install a partition junction stud. You have already been shown how to make one, and you nail up the partition stud as you would any other stud.

All that is left to do on this wall frame at this point is to cut and nail in the cripple studs below the window.

FRAMING THE OTHER WALLS

Brace the first wall frame so that it can resist high winds and other expected forces against it. Be certain, however, that the braces do not interfere with the next wall line,

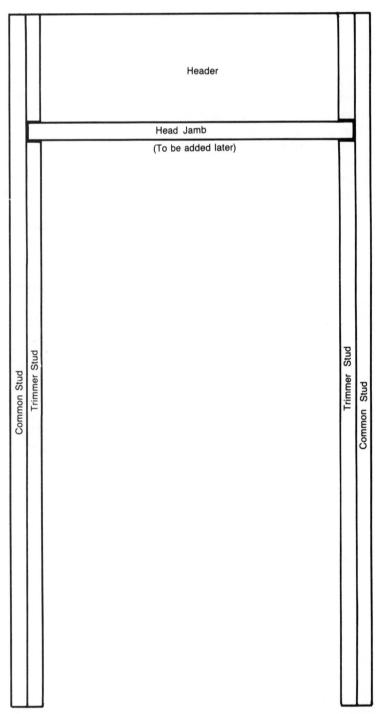

Fig. 11-3. When the trimmer studs are in position, set the header atop the ends of the trimmers and nail the header into its proper position. Make certain it is level.

which will join the first. Before you leave the first wall, check again by using a level and square to see that all four corners of the wall are true and that the wall itself is plumb. If not, make the needed corrections before starting the second wall.

Use the chalkline again and mark a line from the outside edge of the first corner post to the point where the outside edge of the corner post of the next wall will be installed.

Select the timbers for the soleplate and measure and cut these. Then nail them in place, using 20d nails spaced 12 inches apart. When this is done, construct and nail in the next corner post as you did the first ones.

Now proceed exactly as you did before, marking locations for common studs, rough door frames, rough window openings, and any partition junction studs needed. Be sure to check the corner of the soleplate and the first wall line to be sure that the corner is square. If it is not, true it before proceeding to further steps. If the corner is square, proceed to the next steps.

At this point, nail in the remainder of the soleplate the rest of the way around the room, then erect the remaining corner posts. Brace all corner posts securely, and check and double-check for plumb and for square corners.

It is a good idea now to raise the wall framing for the walls that will join the first wall line. This means that you need to use the stud spacers again and nail in all studs, common and partition junction, for the second wall, and then add the top plate for the second wall. This time, allow the top plate to extend 3½ inches past the first stud; that is, past the stud that will be nailed to the first corner post you erected.

The end of the top plate, then, should be flush with the outside edge of the first corner post, and the other end should extend 3½ inches past the last stud so that it, too, is flush with the outside edge of the corner post.

Nail in the top plate, as you did before. Now the two walls are connected at the top of the first corner post you erected, and the end stud will also be nailed to the first corner post.

When you nail the tops of the studs to the top plate, you will use the stud spacers and the C-clamp for the studs close to the first corner post. As the wall proceeds away at a right angle, however, the roof line is high enough to allow studding to be nailed in from the top. It is even possible at this time to assemble the short wall frames and lift them as a unit, if you wish to try it in this fashion.

Complete all rough door and window opening framing. Then proceed to the third wall, which will also be connected to the first wall frame constructed. Complete it as you did the second wall, including the top plate extension so the wall can be nailed to the top of the corner post of the first wall and the first stud can be nailed to the side of the corner post. Make and install all headers, trimmer studs, partition studs, and cripples.

Finally, chalk off the wall line for the fourth wall. Then nail down the soleplate and mark off stud locations, partition junction stud locations, and window and door openings. Complete the installation of the top plate and all studding, and connect the fourth wall to the second and third walls.

Be sure to brace all walls carefully and securely as you complete work on them. Remember to nail 2-×-4 blocks to the floor so you can attach the braces to these blocks.

Do not neglect to check all corners for square and all walls for plumb from time to time. It is possible that you measured incorrectly or a brace slipped as it was being nailed.

Any of the mistakes that can be corrected at this stage should be taken care of immediately. The further you go, the harder it is to make the needed corrections.

BUILDING THE PARTITION WALL

The particular area used in this instance is too large to use as only one room. It can be utilized better if it is partitioned into two rooms.

At this point all of the walls are connected, but the cap has not been installed on any of the walls. There is an excellent reason for this delay: the partition walls cannot be connected to the other walls if there is a solid cap on the top plate. You can, of course, nail in the caps, but be sure to leave space for the top plate of any partition walls.

To start the partition, go to the partition junction stud you installed in the first wall and locate its exact center. Tack a small nail at the center point.

Go to the opposite partition junction stud, locate the center, and mark it clearly. Loop the ring at the end of the chalkline over the nail and stretch the line to the mark at the other partition junction stud. Snap the line so you have a clear line for the center of the next soleplate.

If your chalkline is correct, you should have exactly the same distance from the soleplate of the parallel walls to the line at either corner of the room.

Check it out. Go to one corner and measure from the soleplate to the line. Then go to the other corner of what will be the partitioned room and measure again. If the distance is not the same, the room will not have properly squared corners.

You can now go to the one end of the chalked line, measure off 1¾ inches on each side of the line, and mark the locations clearly. Do the same at the other end. These marks you have just made represent the location of the soleplate that will cross the room area and ultimately become the foundation of the partition wall.

Cut and install the soleplate at this time, and then cut and nail up the top plate of the partition wall. Nail in the end studs so they fit snugly against the partition junction studs, and then add all other common studs. If there is to be a doorway in the partition wall, mark off the rough door opening and frame it as you did before.

The partition wall is now framed, and you can turn your attention to capping the entire frame assembly. First, however, make any last minute checks that need to be made. Check all corners of the partition wall frame to be sure that the room will be square, and recheck the studding where the partition wall frame joins the exterior wall. Make certain that all walls are nailed together tightly and securely.

When you are confident that everything is correct, you can go back to the rough door openings and saw out the portions of the soleplate that were left in the rough openings. After you have removed the soleplate sections, you are ready to cap the wall.

You can easily make the joint of the partition wall to the outside wall much stronger by constructing a T-post. Use a T-post whenever a partition wall meets an outside wall. This can be made in any of several ways. The simplest method of making one is by centering and nailing a 2 × 4 on the face side of a 4 × 6.

Some carpenters prefer to use two 4 × 4s nailed together with a 2 × 4 centered and nailed against the face of the 4 × 4s. Very large nails should be used to nail the 4 × 4s together, and 16d nails can be used for the 2 × 4s.

The most common method of making a T-post is by nailing two 2 × 4s together with blocks between them and a 2 × 4 centered on the wide side and nailed in place. In rare instances, T-posts are made of a 2 × 4 centered and nailed on a 2 × 6. This will then be nailed against bridging that has been nailed between the studs where the T-post will be nailed in place.

The purpose of the T-posts is to provide a substantial nailing surface for the finish work of the inside wall. Do not omit this step, therefore, in your framing. If the partition wall is to be finished on one side only, you can use only a common stud, instead of a T-post.

Partition walls are either bearing walls or nonbearing walls. The bearing partition wall supports the ceiling joists, and the nonbearing wall supports only itself.

If the partition wall is in a one-story building, a common building practice used years ago works extremely well. You can nail in the soleplate but omit the top plate, and the studs are extended far enough to nail to the roof rafters.

This practice has a purpose: the lengthened studs add a great deal of support to the roof, and the studding nailed to the roof creates a much greater stability to the room than would framing that is nailed only to the partition junction studs. You can strengthen the room even more by making certain that a ceiling joist is installed against the studs. Then the studding can be nailed to the joist as well as to rafters. One other argument for the use of extended or lengthened studding is that if you ever decide to finish an attic room for family use, the studding will already be in place.

The usual method of framing works well enough that there is no real need for the lengthened stud approach, unless you happen to have a particular preference or need for it.

BLOCKING AND CAPPING

Before you leave the wall framing, two more important jobs need to be completed. The first of these is blocking, which is especially important beside door and window rough openings.

Blocking is the installation of short horizontal sections of 2 × 4s cut to the exact length so they will fit snugly between studs. This blocking is installed between common studs but not between the cripples.

The reason for the use of blocking is simple: the room frame is strengthened and supported by the vertical studding. You can also now add stability and strength to the vertical support by the use of blocks to keep the studs in place. The stud thereby can do the work for which they were intended.

If a stud is weak or green, it may bend under the weight that is placed upon it, and if it bends, it is not supporting its load adequately. Equally important is the fact that when any straight length of material bends, it becomes slightly shorter. The shortened stud, even if it is shortened only by a fraction of an inch, has pulled partially free from the nails that were intended to hold it in place. This means that the top plate can sag slightly, and your wall frame is no longer totally true.

If you know your studs are strong and completely dried, you can use only one length of blocking between studs. As stated above, cut the block exactly to fit. A block that is too short will not serve its intended purpose, and a block too long will cause the studs on either side to give or bend slightly in order to admit the block. You have therefore again defeated the purpose of the blocking.

There is only one time that an especially long block is useful. If you have a stud that is not perfectly straight or has bent since its installation, you can force it back into its original vertical line by using a block that is slightly too long.

Suppose your studs have all been straight and have been blocked before you reach the bent one. If the bow is away from the line of blocking, omit the blocking on the wide space for the time being. If the curve is toward the line of blocking, you can use a plumb line to determine exactly where the inside surface of the stud should be and cut a piece of blocking to that length. When you start to install it, wedge it in place as far as you can, then, with one end in its proper location, use a hammer to tap the other end into position. Then nail in the block.

When the curve is away from the line of blocking, start at the stud on the other side and install the blocking to force the stud into a straight position. Then cut and nail in the block on the opposite side.

For easier nailing, stagger the line of blocking by the width of a 2 × 4. Nail the first one in place against the corner post by angle-nailing or toe-nailing it. Then fasten the other end by driving a nail through the stud into the end of the blocking lumber. At the next stud, position the blocking piece just slightly above or below the first piece, so that you can nail through the stud instead of having to angle-nail.

If you have reason to believe that your studs are slightly green, you can use two blocks against the corner post, each one-third of the distance from the end of the stud. At the next stud space, install the blocking in the center, exactly halfway between the first two blocks. Continue this pattern the rest of the way along the wall.

The blocking is doubly important on both sides of windows and doors. The blocks should be located at the mid-point of the stud, rather than a third of the way from the top or bottom.

When the blocking is completed, you are ready to *cap* the wall frames. This capping is sometimes called *double-plating*, but all that it amounts to is nailing long 2 × 4s on top of the top plate that you installed earlier.

The cap or cap plate is extremely important, however, because it connects the wall and the roof. It also connects the outside walls with the partition walls.

We have already instructed you to leave a space above all partition junction studs the width of a 2 × 4. To cap the wall frames, start at any corner. You want the cap plate to cross the point where the two walls join. That is, you attached the second wall to the first by nailing the final stud of the second wall to the corner post. Now you need to strengthen the wall even more by bridging across this juncture.

Do not allow joints in the top plate and the cap plate to occur at the same place. Use a longer or shorter cap plate to stagger the joints. Nail the cap plate to the top plate by using 16d nails spaced 12 inches to 15 inches apart.

Cap plate the two end walls first. That is, install cap plate timbers so that wall junctures are bridged.

The top plate of the partition wall fills in the space left earlier. You can cap plate the longer walls now. Nail the cap over the top plate of the long walls as well as over the end of the top plate where the partition wall was joined to the longer walls.

In this particular project, the roof was already in place before the room was built. A space was left at the top of the corner posts—enough room for two 2 × 4s to fit between the corner post and roof joist. At this time the cap plate was slipped into place and angle-nailed, because it is impossible to nail from the top. When the cap plating is completed, so is the wall frame. The wall is now ready for the sheathing.

ONE-PERSON SECTION WALL FRAMING

If you want to try the section method of wall framing, even though you are working alone, you can do so if you have at least average physical strength and if the walls are not very long. If the room to be framed has two long walls and two short ones, it will be possible for you to construct the short walls as units and raise them unassisted.

First, lay out the soleplate and mark it for corner posts and studding, as well as for partition junction studs, if they are needed. Do not nail the soleplate to the subflooring, however. Lay it flat instead, with the stud markings on the upper edge. (See Fig. 11-4.)

You can now lay the top plate alongside the soleplate and mark it similarly for studding, partition junction studs, and corner posts. Mark the top plate by using a straightedge and continuing the lines for the soleplate marking onto the top plate. In this fashion you can be certain that if the soleplate was marked off correctly, the top plate will also be marked in a proper fashion. Set the top plate 8 feet away and lay the studs, corner posts, and partition junction studs or T-posts—whatever will be in the wall frame—in place. (See Fig. 11-5.)

Now stand the soleplate on edge and nail the studs and other timbers into place by driving 16d nails through the soleplate and into the ends of the studs. (See Fig. 11-6.) Place the top plate into position. Attach it by driving more 16d nails through the top side and into the other ends of the corner posts, studs, and partition studs.

You also can frame the rough window and door openings. When the frame unit is complete, lay a short length of 2 × 4 three or four inches from the top plate and parallel with it, and lay several waste blocks of wood nearby. Use a crowbar or prybar now to

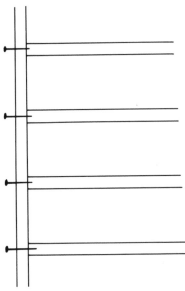

Fig. 11-4. Place the soleplate so the ends of the studs butt into it. You can install the studs by nailing up through the bottom of the soleplate and into the stud ends.

lift the assembly slightly. Insert the blade of the crowbar under the top plate and position the short 2 × 4 so that it can serve as a fulcrum.

Push downward on the crowbar, and when the top portion of the frame assembly is off the floor a few inches, use your toe to push the waste blocks under the lifted frame assembly. Put one block in the center and one at each end of the frame. You now have adequate lifting space so that you can get your hands under the assembly without difficulty.

Fig. 11-5. All parts that will be used in the final wall frame should be laid out for assembly. The studs shown are especially large and are not 16 inches on center; instead, they are 24 inches on center.

Go now to the center of the wall frame assembly and select a stud that is midway from either end. Cut a length of 2 × 4 that is 6 feet and 6 inches long. Using only one nail in the center of the short 2 × 4 about 3 inches from the end, fasten the short 2 × 4 about 3 feet from the top plate. Drive the nail nearly all the way in and leave the other end free. The free end should extend under the soleplate and 12 inches or so beyond. It should swing freely when the wall is lifted.

Measure 6 feet and 4 inches from the chalkline for the soleplate and nail a 12-inch-long block of 2 × 4 parallel to the soleplate. Then nail a 10-foot 2 × 4 timber to the outside of the corner post on one side and 4 inches from the top. Lay the free end of the 10-foot timber over the top of the block you just nailed to the floor. (See Fig. 11-7.)

Now lift the entire wall-frame assembly carefully. As you do, the free end of the short 2 × 4 will drop to the floor. When the wall assembly is chest high, the free end of the short timber will be virtually vertical. (See Fig. 11-8.)

When the timber is vertical, you can rest the wall-frame assembly on the timber while you reposition your hands. Rest briefly before lifting the frame into place. (See Fig. 11-9.)

As you move the wall-frame assembly into a vertical position, the end of the 10-foot timber will slide across the short 2 × 4 nailed to the floor. The timber will be pushed against the 2 × 4 as you allow the wall frame to tilt backward slightly. (See Fig. 11-10.)

Fig. 11-6. Note how the wall frame brace is already attached to the stud.

Nail a brace to the other corner post and run it to a block nailed to the floor at a right angle to the wall frame. Nail the other end of the brace to the short 2 × 4 on the floor. The wall frame is now firmly anchored on one end.

If you nailed the parallel 2-×-4 section with only one nail, you now can tap the short length of wood until it turns from a parallel position to a right angle to the wall frame.

Fig. 11-7. The other end of the brace lies under the top plate so that when the wall frame is lifted the brace will be pulled across the block already nailed to the subflooring.

Now nail the free end of the 10-foot timber to the short segment of wood, and both ends of the wall frame are stabilized. Tack the braces in place until you have a chance to put your level on the posts to see if they are plumb and to make the necessary adjustments. (See Fig. 11-11.)

When the corner posts are plumb, nail the soleplate to the subflooring by using four 20d nails, two side by side in two sections, between all studs. (See Fig. 11-12.) Before you leave, you can add extra braces until the wall frame is joined to other walls.

You can do the other end the same way. When you are finished with it, you can build the long walls piece-by-piece if necessary. Do not attempt to build and lift even short walls if you have any concerns about your back or any other part of your anatomy or health in general.

Fig. 11-8. Lift the wall frame and position yourself so that you can walk forward until the frame is close to its final position. The brace will slide along the floor as you move.

When you are cap plating, use a step ladder that is strong and sturdy. Be certain that all four legs are positioned soundly and that the ladder is balanced adequately.

If you do not feel strong enough to lift the wall assembly alone, you can construct a skeletal wall by assembling soleplate, top plate, corner posts, and every other stud. If this is also too much, assemble only corner posts, soleplate, and top plate. You can nail in the studs and rough window and door frames once the basic assembly is in place.

At no time should you take risks with your body or health in order to save time and money. If your health is good and you wish or need to work alone, however, you can do a great deal of wall framing unassisted. The work progresses more slowly, but you will also take your time and see that everything is done as you wish.

As you finish the wall framing, you can remove the bracing that had been supporting the walls. With cap plating completed, there is no real danger that your wall frames will not stand alone until you can get the structure dried-in.

Fig. 11-9. As the wall frame rises, the brace will fall into position and hold the wall frame in a nearly vertical position until you can nail it in place.

Fig. 11-10. The brace is now pushed against the block by the weight of the wall frame. You can now prepare to nail a second brace when you have the wall vertical.

Fig. 11-11. Now the wall is fairly stable and you can hold or tape a level against the stud. With your hammer in your other hand, you can position the brace and nail it as soon as the wall is vertical.

Fig. 11-12. The hard work is done at this point, and you can nail the soleplate permanently to the floor. Leave the braces up until the wall is safely anchored to other walls.

12

Preparing the Wall for Finishing

WHEN THE WALL FRAMING HAS BEEN COMPLETED, WHEN ALL OF THE
studding is in place and all bracing work has been completed, it is time to prepare the
wall for the finishing touches, whether that includes paneling, wallpaper, siding, clapboard,
or whatever kind of interior and exterior finishing work you plan to include. This chapter
concentrates largely on how one person, working totally unassisted, can handle virtually
any wall-finishing problem that arises.

INSPECTING ELECTRICAL WIRING

In some building codes there are clauses that require a licensed electrician to install
all wiring and switches. These same codes, however, often stipulate that the builder of
the home can do his own wiring without a license; he simply cannot do wiring for other
people.

If you hire an electrician, his work must be inspected by the local building inspector.
If you do it yourself your work also will have to be inspected.

In any event, it is not a bad idea for you to make your own visual examination of
the wiring once it is completed. Even the professional electricians can make mistakes.
When we completed our house, the electricians' work passed inspection. We flipped a
two-way switch, and an explosion as loud and sharp as a rifle shot erupted from inside
the wall. The room was filled with an acrid, smokey smell before the switch could be
turned off.

The explanation was that the wiring had been incorrectly connected. Such a mistake
could easily have led to extreme danger and damage, and possibly to tragedy.

The building inspector typically makes spot checks around the house to see that all
electrical work has been done properly, but often he will not check all of the switches
and sockets. It is your house, and therefore it is your safety that ultimately matters. There
is no reason you shouldn't make your own check.

170

As you are installing insulation, you will have excellent opportunities to see the wiring at close range. While you are in the vicinity of the switches—with the power off—check the wires leading into the switch. Be sure that all of them are tight and that there are no irregularities of any sort. If you find any questionable matters, call the electrician back and have him examine the situation and make the needed repairs.

INSTALLING INSULATION

One of the steps you will want to complete is the installing of insulation. The type and quantity of insulation you choose to use depends greatly upon the climate where you live. Nearly every region in the country, however, has a climate that requires insulation of some type, whether it is to keep out the cold or the heat.

Proper installation of insulation is assuredly one of the most economical steps you can take in the building of a room or house. A high-grade insulation barrier will, in the course of a very short time, repay the cost of the insulation many times over in savings on the heat or air conditioning bills.

Good insulation also helps greatly in the control of moisture inside walls. This barrier against moisture is highly desirable, because the excessive dampness inside walls can cause soleplates and studs to decay at the bottoms and eventually can cause walls to sink or sag badly. Damage can also be done to door frames, windows, and subflooring.

Another excellent quality of insulation is that it will help control fungus growth, which, if unchecked, can damage the inside of walls seriously. In a short time the damage no longer would be confined to interiors but also would cause damage to interior wall finishing and to exterior wall coverings.

One of the greatest advantages of insulation is that many commercial types are fireproof. If you install proper insulation and interior wall coverings, you have created an effective fire wall or fire barrier in the event of a house fire.

Several studies have shown that a wall switch that has loose wires or defective wiring can create a slight spark. This spark can on disastrous occasions find enough dust or wood fibers to keep it alive until it can ignite studding or other fuels. Good insulation can help keep sparking from igniting the interior of the wall.

In walls where there is adequate room, *roll insulation* is usually installed. These rolls are often referred to as *blankets* or *batts*. If there are areas, such as around windows and doors, where there is not sufficient room for batt installation, *insulation pellets* are often used.

When you are working with insulation blankets, dress for the occasion. The insulation fibers, or fiberglass, can cause severe skin and eye irritations, and inhaling the dust from the blankets can cause respiratory difficulties.

Proper dress includes long sleeves, long trousers, a hat or cap, possibly a handkerchief or neckerchief tied around your neck or your shirt buttoned at the collar, and a surgical or other comparable mask covering your nose and mouth. You should also wear gloves that fully protect your hands and wrists. Don't forget your protective eyeglasses or goggles. When you finish work, remove the outer clothing and store it in an appropriate place until it can be properly laundered.

You should never handle the insulation blankets so that the insulation itself touches any part of your body. When you will be working for long periods at a time, you should see that the room is well ventilated, and you should also leave the room at regular intervals to get fresh air.

One side of an insulation batt is covered with paper. There is usually a perforation in the paper every eight feet. If you are insulating between studs, you can tear the insulation along the perforation cleanly and easily, and you have the proper length for the stud areas.

If you have to cut the insulation for irregular installation, you can do so easily with a pair of sharp scissors. Keep all of your tools and equipment together and do not use them for any other purpose until they can be cleaned thoroughly.

The covered or backed side of insulation batts is called a *vapor barrier*, and this insulation should be installed so that the vapor barrier is nearest the warm side of the wall. That is, the vapor barrier should be next to your body as you are working. (See Fig. 12-1.)

If you have spaced your studs properly, the insulation should fit snugly between studs and leave no unprotected areas at the sides. In fact, the insulation will probably be slightly too wide, and you will have to force it lightly into the space between the studs. This tight fit is excellent; do not attempt to trim off the excess on the edges. Some rolls of insulation are marked in feet and inches along the vapor barrier, so you don't have to measure each time you need a special length.

On each side of the insulation batts there are flanges or lips that spread slightly to the sides. These flanges are for nailing or stapling the insulation in place.

Any time you must cut a batt shorter than the manufactured markings, save the shorter piece and lay it aside. Each time you have excess length, add the next piece to your stack. Later, when you are insulating around pipes or other tight places, you can use these shorter scraps and not have to cut a full-length batt.

If you are working with unmarked batts, start with the space between pairs of studs and cut the batts 8 feet and 3 inches long. The extra length will provide you with nailing or stapling space at the top and bottom of the batt.

Do not try to save money by cutting the batts for a perfect fit. An amazing amount of hot air in summer and cold air in winter can enter a house via the small spaces left at tops and bottoms of insulation batts, and trying to save a few inches of insulation is false economy.

When you measure the batts, use sharp scissors to cut along a straight line. Then lift the batt by holding the end so that the actual insulation is away from you and the vapor barrier is nearer your body.

Spread the nailing flanges and push the insulation batt into the space between studs. As you do so, fluff the batt lightly for maximum protection. When you push the blanket into the space, it will fit tightly enough that it will stay for a moment or two without support. During this time get either your hammer or staple gun and prepare for permanent installation.

Whether you use a hammer and nails or a staple gun is purely a matter of your own personal preference. We have found, however, that the use of a staple gun is much easier and faster.

If you use nails, be sure to use those with wide heads so that the head will not allow the flange paper to slip over it. You can buy a special type of nail or tack, similar to a roofing nail, that works very well.

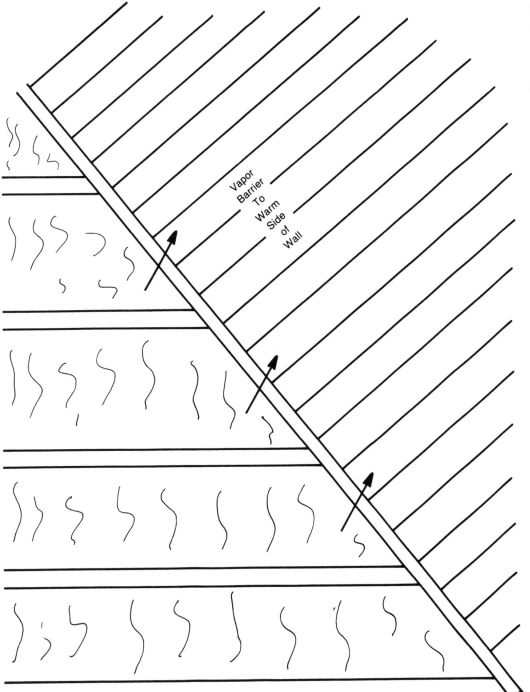

Fig. 12-1. Insulation should be installed so that the vapor barrier faces the inside or the warm side of the room. The vapor barrier is ineffective if installed the wrong way.

Vapor
Barrier
To
Warm
Side
of
Wall

If you use a staple gun, all you need to do is spread the flange so that it lies against the edge of the stud. Then hold the staple gun against the flange and begin stapling. Do not try to save money by using one staple every 3 or 4 feet. Staple every 6 to 8 inches.

Start at the top of the blanket and staple a foot or so down one side, then move to the other side and staple down the same distance. Move back and forth across the blanket as you progress from top to bottom, until you have stapled the bottom in place. Then go back to the top and put a row of staples across the very top so that all air penetration is eliminated, as far as is possible.

You can also start at the bottom, but many people find it inconvenient to do so. The blanket is in your way as you work, or it is draped awkwardly over your shoulders and back.

If you are working around pipes, be sure to pack as much insulation as you can behind and around the pipes, particularly on the cold side—or the side near the outside wall. If you live in a cold climate, failure to insulate on the cold side of pipes can lead to frozen and broken pipes and a great deal of interior wall damage by water. Many people also insulate cold water pipes to prevent condensation and subsequent dripping on hot days.

After you have insulated all of the full stud spaces on the exterior wall, go back and fill in all of the irregular spaces, such as those where bracing and blocking have been installed. In stud spaces crossed by diagonal bracing, you will need to cut the insulation to fit the spaces. The simple method of doing this is to measure from the top of the stud space to the top of the brace. Then measure on the other side of the stud space to the bottom of the brace. Measure not just to the brace but to the other side of it, so that your blanket will be about 2 inches longer than the space.

When you get ready to cut, put a mark for the top and bottom measurements and cut diagonally across the blanket. Use the excess length to pack the space near the brace tightly. You will need to do this above and below all braces.

To insulate around blocking, measure to the blocking and add 3 inches. Cut the blanket straight across and install as you would for a full stud space. (See Fig. 12-2.)

When you have completed insulation around braces and blocks, inspect the rest of the wall for any spaces you have neglected. You should pack insulation into any cracks that will contain even small amounts of insulation. It is incredible how much cold air can enter a house around the switch box near a door, and the same is true for the spaces around windows and doors.

When you have completed all walls, you will probably have a small pile of scraps left from your work. Go back around the walls and find any spaces not insulated fully. Even if you cannot find a deficient spot, pack the scraps in beside regular batting for extra protection against the elements.

Here is an example of how much heat and cold loss can be prevented by insulation. An insulation blanket that has the usual $3\frac{5}{8}$-inch thickness has an R value of 14.24. (The higher the R value, the higher the batts insulating ability.) An 8-inch-thick log wall has an R value of 8.10. By comparison, the insulation blanket is nearly twice as effective as the log wall. A 12-inch log wall has an R value of 12.07, still not as great as the $3\frac{5}{8}$ thickness of an insulation blanket.

If you use wall coverings attached to some form of lath, there is a chance that the lath pattern will show through many types of walls (except plywood) if there is not an adequate vapor barrier behind the wall covering. Good insulation includes the vapor barrier on the insulation batts.

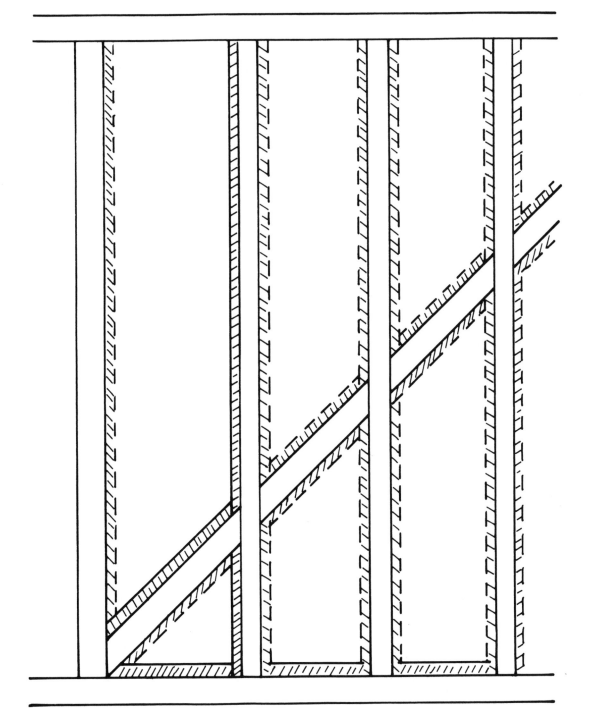

Fig. 12-2. Be sure to cut insulation to fit into spaces above and below blocking and bracing. Cut the sections long enough so that you will have enough material to nail or staple to the braces and blocking.

Insulation is so important that it has been demonstrated that steel structure buildings tend to collapse under great heat if the steel is not protected by a special insulation.

Some builders have found that the installation of polyethylene sheets is effective as an additional vapor barrier. This is more particularly used in ceiling installation under ceiling joists. The polyethylene is installed before the furring strips that support ceiling tiles are nailed in place.

You can also help cool and heat your house if you insulate between floor joists under the house. An immense amount of air can penetrate the subflooring and flooring, adding to the cost of heating or cooling your house.

PREPARING WALLS FOR COVERING

When the insulation installation is completed, you may wish to start preparing the wall for the covering of your choice: paneling, boards or planks, or other forms of wall surfaces. Most of the commercially prepared panel coverings are manufactured in sheets 4 feet wide and 8 feet long. Most of these can be nailed up or glued in place with a minimum of difficulty.

You can make the job easier by preparing the stud surface in a few simple ways. Start by making a simple test to see if the stud edges are all uniform at all points. Go to one corner of the room and tap a finishing nail just deep enough that it will withstand a slight pressure. Drive the nail partially in at the very top of the corner post. Place another nail in the corresponding place at the other corner post for that wall.

Tie a length of small cord around the first nail. Push the cord down so it is barely missing the surface of the corner post. Then stretch the cord to the other nail and do the same thing. Be sure the cord is taut.

Now move across the length of the room and examine the point where the cord crosses each stud. There should be the same amount of clearance at every stud. You might find, however, that the cord actually passes very close to or far from a stud. In this case the studs do not provide a surface level enough to support wall covering.

If a stud sticks out too far, you might be able to hit it sharply with a hammer at the top and bottom and drive it back into place. If not, you need to remove the stud and renail it correctly. If this does not solve the problem you need to shape or reduce the stud by sanding or planing its surface until it fits in the wall at the correct level.

You might find that one or more studs had been recessed too deeply, with the result that the cord missed the stud by as much as $\frac{1}{4}$ inch or $\frac{1}{2}$ inch. You need to remove the stud and install it correctly or locate a stud that is the proper size for the space.

If no corrections are made, you will find that when you start to nail up plasterboard, paneling, or pine boards, the edges of these materials will not be flush. The irregularity will also create difficulties later with the installation of ceiling and floor molding.

If the cord indicates proper stud placement, move the nails down to the halfway point and check again. Do the same near the bottom of the studs.

Correcting placement problems simplifies the installation of plywood and plasterboard. Otherwise the fact that the edges are not flush will be readily apparent, even prominently so each time you look at the wall.

When the nailing surfaces of all studs are satisfactory, your next step depends upon the type of wall covering you have chosen. If you are setting wide pine or other types

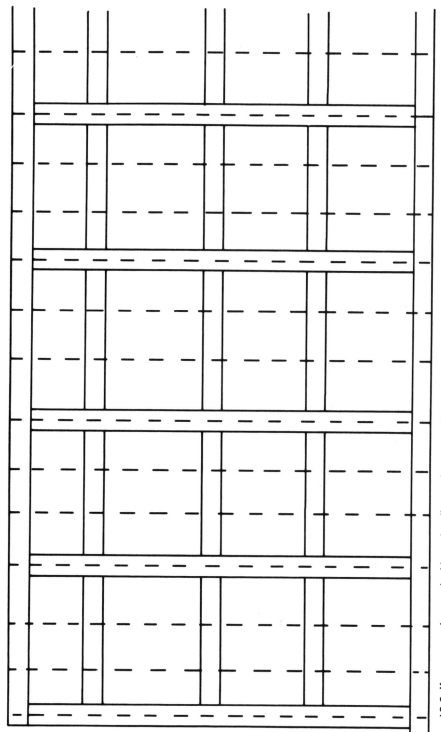

Fig. 12-3. If you are using vertical board wall covering, you will need at least three blocking sections in addition to the top cap and soleplate as nailing surfaces.

of boards vertically, you will need to install additional blocking as support for the boards and to provide a nailing background. (See Fig. 12-3.)

Cut 2-×-4 lengths that will fit snugly between studs. Nail these into place one-third of the way down and two-thirds of the distance to the floor. When you nail the boards in place, you can nail them to the top plate, to the two blocking pieces, and to the soleplate. If you plan to end the boards at the top of the baseboard, however, you will need one blocking piece at the level for the top of the baseboard. Because the boards come from the supply house in 8-foot lengths, however, it is pointless to cut off the final 6 inches to end them atop the baseboard. It might be much easier to run them all the way to the floor and use a narrow baseboard or only floor molding.

If you do not use the blocking lengths, you will have only the top plate and soleplate for nailing surfaces. The boards will have a tendency to give, even if you use thicker lumber, in the sections between the two plates. This giving motion can be damaging to tongue-and-groove lumber. If the lumber has a regular edge, there can be enough space created to permit small insects to enter your room.

You can use a diagonal pattern for board wall coverings without the blocking pieces, since each board will cross several studs. Nailing the boards to the studs can provide adequate support. (See Fig. 12-4.) These blocking lengths are called *nailers*.

You can attach some types of wall panels more effectively by nailing or glueing them to a furring strip or lath attached to the studs. This method is recommended because the furring strips or lath will reduce the joint movements. This movement is created when the studs and nailers swell or shrink because of the temperature and the humidity inside the room. You will find that this method of installation is quite easy.

To install furring strips, cut or buy 2-inch furring strips and nail them in place at regular intervals from the top to the bottom of the studding. When you are ready to install the plywood you can apply glue to the furring strips and put the plywood panels in place. Then nail them with finishing nails spaced every six inches along the studding to provide additional stability.

Be sure to remove all nails used for cords or other temporary purposes. Check all stud surfaces for any errant staples or other objects that will cause damage or create difficulties when you start to install the wall covering.

Clean the work area thoroughly before you begin covering the walls, because grit, nails, and other unwanted objects on the floor will scratch surfaces of plasterboard, paneling, plywood and other forms of wall covering. Dirt, dust, and other debris that is not cleaned up will find its way, in part, into the space near the soleplate. Dust, along with sawdust and wood fibers, can become a serious fire hazard. A good sweeping or a little time with a vacuum cleaner, however, will eliminate the problem. Dirt on the floor can cause unsightly or permanent spots on the paneling or wallboard, even if the dirt does not constitute a damaging factor otherwise.

PREPARING FOR SIDING INSTALLATION

Siding comes in a wide range of types and materials. Some of it is guaranteed not to chip, check, or need painting for at least ten years. Some siding is simply wooden boards, usually 1 × 5 or 1 × 6, nailed horizontally across the 2-×-4 studding. You can buy siding that looks like cedar shakes, rough-cut boards, bricks, or several other styles.

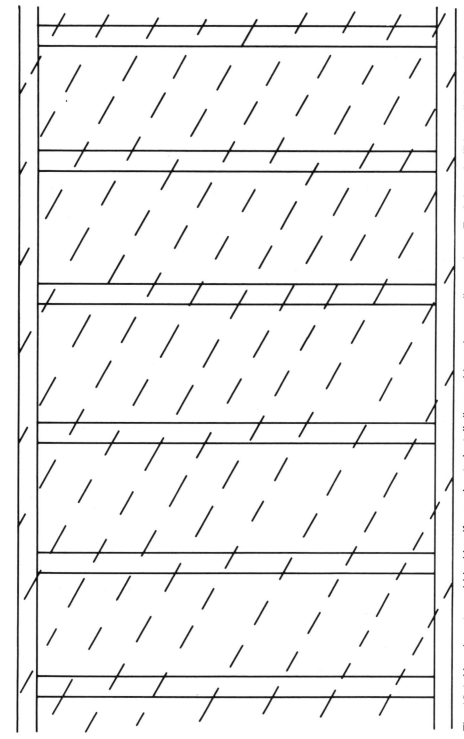

Fig. 12-4. You do not need blocking if you plan to install diagonal boards as wall coverings. Each board will be nailed to at least one stud as well as to the soleplate and top plate. Some boards will be nailed to as many as five or six studs.

179

To prepare the outside walls for installation of siding, you need to make some of the same checks you made for installing wall covering indoors. The first of these is to check for alignment of studs. You can do this by using the cord and nails as you did for the interior walls.

If the house is several years old, check also for signs of damage to the studs by termites, fungus decay, or water, particularly at the bottoms. If a stud is badly damaged, take it out and replace it. Many do-it-yourselfers attempt to find the easy way to solve a problem. Often an amateur carpenter leaves a stud with a decayed bottom in the wall and simply nails another length of 2 × 4 alongside it. This practice is not advisable, because the rot or decay is still present, and there is little to keep it from spreading up the stud. More importantly, the stud is supporting nothing, not even itself. It is being held in place only by the wall and by the nails in the top plate.

A stud might be too deeply recessed, either because it was not milled to full dimensions or because of careless installation. You can avoid the problem of removing it and reinstalling it by nailing a sound stud to the side of the recessed one. This time, however, pull the edge of the new stud out slightly so that it is aligned with the others.

You can also hold a long 2 × 4 horizontally so that its edge is in contact with the studding. You can look along the stud line to see if good contact is being made or if some of the studs are so far out of line as to cause the siding to fit improperly.

Make a visual check to see that there are no nail ends or heads that inadvertently have been left protruding from studs. These will prevent the siding from fitting snugly against the stud line. They should be removed before you proceed.

Before you start nailing up siding, you should take the necessary precautions against improper storage of the siding materials. Most of the asbestos-cement types of siding can be very easily damaged by water. Store the siding in a dry room, therefore, and see that moisture does not get to it. It can be water-damaged even if it is still packaged.

Store the siding so that it lies flat and evenly, and see that it is elevated off the floor by at least 5 or 6 inches. You can make rows of 2 × 4s laid three in a stack, with each row no more than 2 feet from its neighbor. Then you can stack the siding packages no more than half a dozen packages high.

Until you are ready to use the siding, keep it stored in the packages it came in. The same precautions should be observed regarding plywood, fiberboard, and gypsumboard.

Prior to nailing up the first siding, take steps to see that water does not run down the face of foundation walls. Such water flow will eventually cause seepage. The result can be infestation by insects, particularly termites and cockroaches, as well as mildew and fungus growth. You can prevent this water seepage in a large part by the installation of what often is referred to as a *water table*. (Such a device is not to be confused with the water tables associated with bored wells.)

At the bottom of the wall frame, nail up a 1-×-4 or 1-×-5 board horizontally so it will lap over the foundation wall by at least 1 inch, perhaps 2. At the top edge of the board nail in a thin beveled or curved strip of wood called a *drip cap*. Water that runs down the wall siding will drop off the drip cap and fall to the ground an inch or so from the face of the foundation wall. (See Fig. 12-5.)

You should also buy metal flashing to nail over the drip cap before you finish the job. At this time the wall frame includes the studding, the sheathing nailed over the studs,

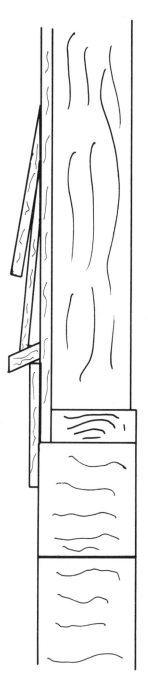

Fig. 12-5. A water table can be installed to prevent water from running down the foundation wall or seeping into the inside of the wall.

the horizontal board at the bottom of the frame nailed to the soleplate, the drip cap installed on top of the horizontal board, and the metal flashing. Such metal flashing should also be installed on the drip cap of windows and doors to prevent water from running behind the door and window frames.

After drip caps are in place at the bottom of the wall, you can nail up starting boards or you can let the drip cap serve as a starting board. The purpose of the starting board is to tilt the bottom of the siding board outward slightly so that water will drip off the edge rather than run down the wall. This starting board can be as thick as your siding and generally no wider than 2 inches.

The bottom of the first siding board will rest over the starting board and thus be tilted out for the drip effect. All subsequent siding board bottoms will rest on the top of the first siding board, and the thickness of the board will create the drip-cap effect.

You will also want to apply some type of asphalt-saturated felt paper over the sheathing on the exterior walls. The felt paper is available in rolls for easy handling. You can use a measure to determine how much you need. If you prefer, you can pull enough paper from the roll to reach to the top of the wall, staple or nail it in place, and then cut it when you reach the bottom.

You can start at either corner and attach the paper to the wall frame over the sheathing. At the end of each strip you can cut the paper easily with a pocket knife or scissors. For the next row lap the felt over the first row by at least 2 inches and preferably 3 or 4 inches. Follow this simple pattern until the entire sheathed surface is covered. If you have a small amount of the building paper left over, cut it into foot-long strips. Nail or staple them vertically over the corners so there is no possibility of water passage in this area.

You can use small nails with large heads, like those used for roofing purposes, to install the building paper. Staples from an ordinary staple gun are fine if you choose to use a stapler.

You should also have the corner covering installed before the siding work is started. The easiest method of corner covering is to use lumber 1⅛ inches thick. For each corner cut one strip of lumber 3 inches wide, and another strip 4 inches wide. Nail up the 3-inch-wide strip first. Cut it to length and nail it flush with the corner. Use 16d nails, two per row, in rows 15 inches apart.

Next, nail up the 4-inch-wide strip of covering stock. Nail it so that the corner edge is flush with the edge of the 3-inch-wide strip. Use the same nailing pattern. When you are finished with the corners, you will be ready for the installation of the siding.

At this point your interior walls should be completely framed, with insulation installed between the studs and in all other points where air passage could occur easily. If you are using plywood or a similar wall covering, you could also have the nailers installed.

We recommend that, if you are using plywood, you should apply a thin bead of wood glue to the nailers, and then position the plywood. Finally, nail the plywood in place with finishing nails just long enough to penetrate the plywood and enter the studs sufficiently to hold the plywood sturdily in place.

On the outside walls you should have installed the sheathing and covered it with building paper. Corner covering should also be completed, and the water table and starting strip should be in place. When you have completed all of this, your walls have been framed and you are ready to install the wall covering inside and outside.

FINAL CONSIDERATIONS

You might want to consider making some of the timesaving and energy-saving devices that can be assembled at virtually no real cost and within a matter of minutes. These simple ideas can help speed up your work and help also to eliminate thoughtless errors that can be so costly and irritating in your work.

For instance, if you need an especially long 2 × 4 for a top plate and you cut it incorrectly, you have perhaps ruined a 20-foot timber, or at best you have made it necessary to use it for some shorter purpose, such as studding. Even if the timber can still be used efficiently, you have experienced the frustration and time loss associated with the mistake.

Do not be discouraged or intimidated by mistakes. No skilled carpenter ever reached his level of accomplishment without making mistakes. Far more important than the mistakes is the growth you can realize in both the areas of accomplishment and self-confidence as you continue to work.

The experiences you gain in wall framing will be of great benefit to you when you undertake the projects involved with wall construction. With wall construction you can see your work taking a finished professional effect very quickly. At this point you are ready to begin making plans to complete the wall by installing interior and exterior covering, both of which are among the most rewarding aspects of home construction or home improvement.

Index